Robert Druitt

Report on the Cheap Wines from France, Germany, Italy, Austria, Greece, Hungary, and Australia

Their Use in Diet Medicine

Robert Druitt

Report on the Cheap Wines from France, Germany, Italy, Austria, Greece, Hungary, and Australia
Their Use in Diet Medicine

ISBN/EAN: 9783743435476

Printed in Europe, USA, Canada, Australia, Japan

Cover: Foto ©berggeist007 / pixelio.de

Manufactured and distributed by brebook publishing software (www.brebook.com)

Robert Druitt

Report on the Cheap Wines from France, Germany, Italy, Austria, Greece, Hungary, and Australia

REPORT

ON THE

CHEAP WINES

FROM

FRANCE, GERMANY, ITALY, AUSTRIA, GREECE,
HUNGARY, AND AUSTRALIA:

THEIR USE IN DIET AND MEDICINE.

BY

ROBERT DRUITT.

SECOND EDITION, REWRITTEN AND ENLARGED.

HENRY RENSHAW,
356, STRAND, LONDON.
1873.

LONDON:
SAVILL, EDWARDS AND CO., PRINTERS, CHANDOS STREET,
COVENT GARDEN.

JAMES L. DENMAN,
20, PICCADILLY, W.,

SOLICITS ATTENTION TO THE FOLLOWING WINES:—

FRENCH.

BORDEAUX RED WINES.

Medoc	14s. per doz.
Claret	20s., 24s., 36s., 42s., 48s., 54s.
Superior 60s., 72s. Ditto, 1st growth	84s., 96s.

BORDEAUX WHITE WINES.

Sauterne	18s., 20s., 24s., 30s., 36s., 42s., 48s.
Superior	54s., 60s., 72s.
Chateau D'Yquem, 1st growth	84s., 96s.

BURGUNDY RED WINES.

Maçon ... 18s. Beaune, Nuits	24s., 36s., 48s., 54s.
Clos-Vougeot	60s., 72s., 84s.

BURGUNDY WHITE WINES.

Chablis ... 18s. Meursault	24s., 36s., 42s., 48s., 60s.
Montrachet	54s., 60s., 72s., 84s.

CHAMPAGNE.

Epernay (recommended)	30s.	Moët's or Clicquot's, 1st quality,	72s.
Do. Superior	36s.	Fleur de Sillery	54s.
Do.	42s., 48s.	Creme de Bouzy	72s.

SOUTH OF FRANCE WINES.

Roussillon, or French Port	18s.	Lunel	42s.
Masdeu	22s.	Frontignan	48s.

GERMAN WINES.

Hock, Still ... { 18s., 20s, 24s., 34s., 38s., 48s., 54s., 66s., 72s.	Moselle, Still ... 18s., 20s., 24s., 36s., 42s.
Do. Sparkling 36s., 42s., 48s, 54s., 60s., 72s.	Do. Sparkling, 36s., 42s., 48s., 54s., 60s., 72s.
	Red Hock ... 24s., 36s., 48s., 54s.

PORTUGAL WINES.

Genuine Alto-Douro, stout and useful }	24s.
Do. rich, full-flavoured, excellent for bottling or present use }	30s.
Do. soft, matured, with character	34s.
Do. rich, with great body	38s.
Old Bottled Port, } 38s., 42s., 48s., rich or dry } 54s., 60s., 72s., 84s.	

SPANISH WINES.

Port, Catalonian	18s.
Do. superior	22s.
Sherry, Arragonese	18s.
Do. excellent	22s.
Do. Cadiz	24s.
Do. superior do. }	30s., 34s., 38s., 42s., 48s., 54s.

Any of the above in Pints 4s. per Two Dozen extra.

Detailed Priced List of all other Wines, Spirits, and Liqueurs, post free on application.

TERMS—CASH.

Country Orders must contain a remittance. To ensure safety, all Cheques should be crossed "National Bank." Post-office Orders to be payable at the Chief Office, E.C.

BOTTLES AND CASES TO BE RETURNED OR PAID FOR.

Single Bottles of Wines and Spirits forwarded.

"What d'ye *[illegible]*"

GLASSES, TUMBLERS, &c., ADAPTED FOR DRINKING THE PURE WINES,
GOOD IN FORM, LIGHTNESS, STRENGTH.

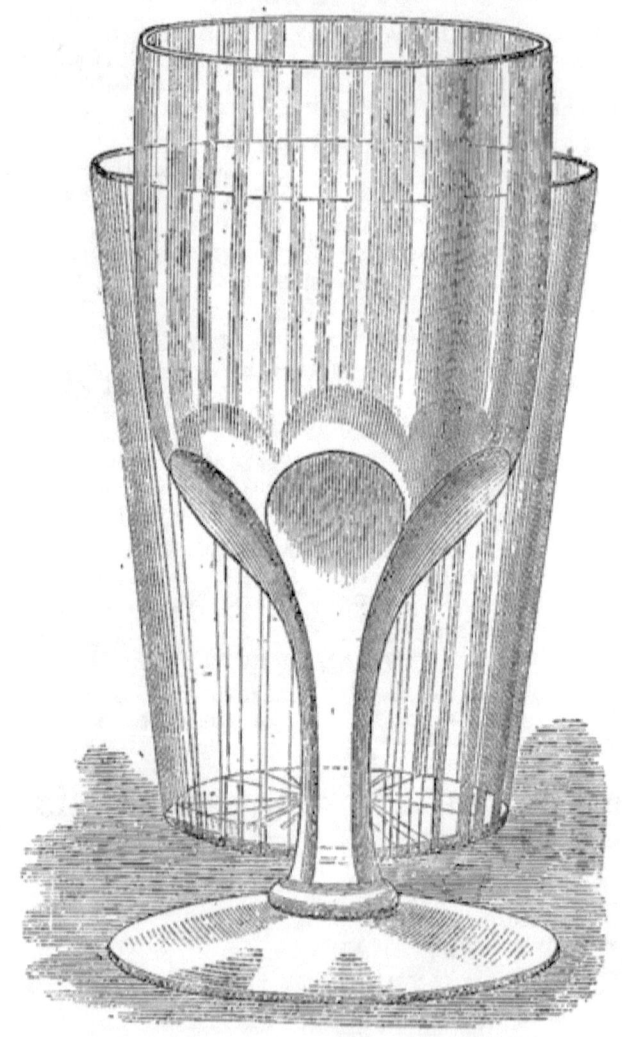

BOUCHER, GUY AND Co.,
GLASS MANUFACTURERS &c., TO HER MAJESTY
GENERAL FURNISHERS IN CHINA AND STONEWARE,
At Prices most moderate,
128, Leadenhall Street, London, E.C.

PREFACE.

THIS is meant for a second edition of a "Report on Cheap Wines" which originally appeared as a series of articles in the *Medical Times and Gazette* in 1863 and 1864, and was published in a collected form in 1865. At that time rivers of strange wines were coming in from all parts of the world, and both the medical profession and the public wanted to know what they were good for. My little book seemed to give satisfactory information on that point; anyhow, it sold furiously and was soon out of print. Scarce a day has passed since in which it has not been asked for at Mr. Renshaw's, and it has brought me the flattering approval of men of letters, physicians, diplomatists, and œnologues. Daily occupations of an absorbing kind and wretched health have prevented me hitherto from revising the book, but I have never lost sight of the subject, and now at last have plucked up heart to make a second edition.

In this not much, of the original work remains. I have not felt it necessary to give at so great length the composition of cheap fortified wine, or to denounce the wine forgeries of Hamburgh. But I have first tried to give such an abstract of wine lore as shall teach any intelligent person what good wine is and ought to be. I have taken pains to show—so far as the poverty of language enables me—the effects of wine on the four senses of touch, taste, smell, and that obscure sense which teaches us the difference between feeling well and ill, and how each ought to have its due share in tasting wine. Now that most houses contain a microscope, it has occurred to me that whilst the young people amuse themselves with the protozoa in ditch water, their elders may study the deposits in bottled wine, and so may be able to distinguish those which betoken disease of the precious liquid. To aid in this study there is a woodcut representing the deposits good and bad found in a great variety of wine this August, 1872. Let me here say that the English are so much in the habit of using fortified wine, which will keep almost any time, that they are apt to keep pure Burgundies and Clarets long beyond the time when they have arrived at their prime, and after which, as Hippocrates said, as they cannot get better they must get worse. Dives therefore should weed his cellar from time to time, and send his surplus to Lazarus—or let me say, not to Lazarus who trades upon his ulcers, nor the institution-mongers who trade upon Lazarus, but to "genteel poverty," widows of limited income, girls at cheap boarding-schools, and homes where "recollections of better days" do duty for substantial food and bright fires.

Next I treat of the wine of each country, its characters, uses, and cost. I do not profess to give "universal information," but only what is to be gathered by personal observation of wines that are to be got here in England. When one considers how many these are and how cheap, one wonders at the *gulicidal* policy (if I may coin a word) which makes some people confine themselves to port and sherry. A cask or a few dozens of light ordinary wine, as Fronsac, Beaune, or Ofner, some samples of good Claret, Burgundy, Naussa, Vöslauer, or Auldana, and a few of white to match, as dry Ruszte, Hochheimer, Montrachet, or white Bukkulla, should form part of the cellar of everybody that can afford it and desires to keep out of the doctor's hands. A sparkling wine, too, is often a true ἰατρεῖον ψυχῆς, or medicine of the soul. Whilst writing this preface I have rummaged out a secret *cache* or hoard of some samples I had put by, seven years ago, when writing my former edition. As to the Somlau and white Ruszte, the St. Elie, and white Keffesia, and the white Diasi and white Tokay of M. Diosy, I can but express my admiration of their exquisite condition, softness, and abundance of flavour. Some sweet wines, too, as Thera, Calliste, and Lachryma Christi, and Boutza are in capital condition. Of the wines of America I know nothing, but witness with delight the effort to make them take the place of the firebrand whisky, and so to deliver that great and energetic people from the incubus of intemperance.*

Throughout my book I have kept constantly in view the difference between pure wine and distilled spirit, and have shown how the addition of the latter to wine affects the sense of touch, and deranges the true sense of taste, all over the gustatory apparatus and beyond. From personal experience, I say that the hot, strong, so-called Sherries supplied at many hotels, and used in many families, are disgraceful to those who sell and dangerous to those who drink them. I do not join in the sensational outcry about drawing-room alcoholism; but say, as a simple fact, that Sherries are often given to women which are too strong to be safe.

In treating of the fortified wines, no man of sense need hesitate to express his admiration for the superb wine flavours of old Port, Sherry, Madeira, and sometimes of Marsala. These are cordials which the wise will use in their proper place. But it is the heavily fortified new wines—which have no flavour—which never will have flavour, because their acids are got rid

---

* See "The New West, or California in 1867-68," by Charles Loring Brace. London: Trübner, 1869. "The Cultivation of the Native Grape, and Manufacture of American Wines," by George Husman. New York: Woodward, 1868. (Mr. Husman unfortunately advocates the system of making sham wine by fermenting sugar and water with the pressed husks of grapes). "Three Seasons in European Vineyards," by William J. Flagg. New York: Harper, 1869.

of by plaster, that I denounce when foisted into daily use. Alcohol is a drug, it costs 2s. a gallon, and though an essential ingredient in wine, is not that which the physician or the connoisseur values. One mouthful of Champagne or old Madeira or Burgundy will produce an instant exhilaration, because of the wine flavours, which no quantity of sheer alcohol and water could produce.

My pages are limited, and so it is that just as I warm with the subject, and long to go into the moral properties of wine, its real value as a nutrient, and the secret of its power, I am obliged to leave off. If my life is spared I hope to come out on this part of the subject, on which I have been collecting materials for years. The only way of dealing with it is the "empirical," or the collection of ascertained facts as they occur in daily life. This is the true scientific basis of diet and therapeutics. Science has no means of dealing with them *à priori*. We know that wine is greedy of oxygen, and that the animal body is an oxydizing machine, par excellence. We therefore affirm that wine is a true food. Parkes teaches us that it increases the action of the heart. Beyond that we know nothing, save by ordinary observation. When the balance and the test tube can detect the difference between a happy and an unhappy man, then chemistry may explain the effects of wine, but not till then.

One word on the wine duties. Some personages

desire that wines loaded with 20 per cent. of proof spirit shall come in at equal rate of duty with wines that are pure—*i.e.*, that spirit mixed with wine shall come in duty free. Moreover, that the duty on wine shall be raised to 1*s*. 6*d*. Some wine merchants seem to favour this, because supposed to promote their interests as against the public. Some authorities also suppose that it will promote trade with the Portuguese. Of questions of finance it is not my place to speak. But let it be said that there is not an honourable and conscientious winegrower on earth who does not deprecate it as a premium on the fabrication of crude, unwholesome, slovenly, ill-keeping wines. Ask MM. Jules Guyot and Terrel des Chênes—ask Korizmics, Schlumberger, and Wyndham of Dalwood! All members of the medical profession, and friends of temperance and morality, protest against it likewise.

R. D.

LONDON, *Oct.* 24*th*, 1872.

☞ The writer begs to intimate to any of his friends, who may not have been directly informed of it, that he is driven by health to pass this winter in a warm climate, and has left his residence in London. Communications may be addressed to him at No. 3, Little Stanhope Street, Mayfair.

CONTENTS.

CHAPTER I.

Object of the book: vast scope of wine knowledge—Importance of it to the medical practitioner—Use of wine as a restorative—Wine often better than tea—Promotes sobriety and good morals—Empirical philosophy of wine—Knowledge of wine ought to be cultivated by all who buy and drink it   pp. 1—6

CHAPTER II.

What is wine?—Varieties of grapes:—strong grapes and weak grapes—Anatomy of the fruit—Ingredients, saccharine and nitrogenous—Fermentation: process of; theories of, the chemical, the chemico-vital, or panspermatistic, and the archebiological—The new wine: the care it requires: racking, sulphuring, and fining. . . . . pp. 7—19

CHAPTER III.

Wine as a subject of chemical analysis—Its proximate principles—Alcohol and its varieties: aldehyde, ethers, acids, earthy salts, glycerine—Abundant sources of alcohol—"Strong drink" of the English Bible—Percentages of ingredients . pp. 20—27

CHAPTER IV.

Wine and its effect on the nerves of taste—Taste, and smell, and touch—Specimens of wine tasted—A light white Bordeaux wine; a light red St. Estèphe; a fortified red wine from L'Hérault; Château Lafite, 1851; Château Yquem, 1858; St. Elie—Varieties of flavour—Rules for tasting wine—Alcohol affects the nerves of touch—Vocabulary of wine terms—Difficulty of describing sensations . . . . pp. 28—42

CHAPTER V.

Importance of spirit to the revenue—Natural proportion of spirit in pure wines of all countries—Mulder's belief in natural port—Government inquiries and reports—Why spirit is added to wine—*Le vinage* in the south of France . . . pp. 43—50

CHAPTER VI.

On the acidity of wine—Diseases of wine; M. Pasteur's treatment—Deposits in bottled wine—Beeswing, crystals, parasitic growths . . . . pp. 51—62

CHAPTER VII.

Classification of wines—Light or pure wine—Sweet wine—Sparkling wine—*Vins de liqueur* or fortified wines—Geographical classification—Wines of France—Political finance and prohibition—Bordeaux and the poet Ausonius—Classification of Bordeaux wines—Médoc—Graves—Petits Graves—Sauternes—Libourne—St. Emilion—Bourg—Blaye—Entre deux mers—White wines—Red wines—Nomenclature of wine—Medical uses—Different kinds of thirst . . . . pp. 63—86

CHAPTER VIII.

The praise of Burgundy: its perfume, its place at dinner—A visit to the Côte d'Or—La Fontaine Couverte—"Saintenay, *cinquante-six*"—Volnay, Montrachet, Mersault blanc, and "l'ingrat midshipman"—Clos de Vougeot, Beaune, Beaujolais, Macon, and Hermitage . . . . pp. 97—100

CHAPTER IX.

On some wines of the south of France—Fine vintages—Good ordinary wines of La Gauphine—Béziers—Château neuf du Pape—Roussillon—Lamalgue . . . . . pp. 101—105

CHAPTER X.

Wines of the Moselle and Rhine—"Scharzhoffberger, 1842"—Fictitious Moselle—Excellence of Rhenish wine—First-class growths—Not used as it ought to be—Swiss wine—Yvorne
pp. 106—111

CHAPTER XI.

Wines of Italy—White Capri—Chianti—Montepulciano—Wine at Sorrento—A perfect hotel—Wine list from Florence—Wines of the Ancients—Vino Rosso Romano . . . pp. 112—116

CHAPTER XII.

Wines of Greece—Old St. Elie—White and Red Hymettus and Keffesia—Red Santorin—Thera—Patras—Naussa—Sweet wines—Le Roi des Montagnes . . . . . pp. 117—121

CHAPTER XIII.

Hungarian wine; well known in the seventeenth and eighteenth centuries in England; praised by F. Hoffmann—A visit to Pesth—The Moslem wave beaten back—Tokay—A digression on sweet wines in general; sweet Ruszte, Cyprus, Malmsey, grape syrup, Constantia, Lunel, Rivas Altes—Syrian wine from Mount Lebanon—Australian Cyprus . . pp. 122—131

CHAPTER XIV.

Hungarian Wines continued—Dry White Ruszte, Szamorodny—Dioszeger Bakator—Œdenburg—Somlo—Neszmely—Badasconyer—Red Ofner, Szegzard, Erlaure, Carlovitz, Visonta—Gold label Ofner—Transylvanian wine . . pp. 132—141

CHAPTER XV.

Austrian Vöslauer wines of M. Schlumberger . . pp. 142—144

CHAPTER XVI.

A digression on mead or metheglin, with a few words on cider—The *Sicera*, or *Strong Drink* of the Bible—Decay of housewifery—Cases in which cider should be prescribed 145—150

CHAPTER XVII.

Wine from Australia—Mr. Patrick Auld's Auldana—Mr. Wyndham's Bukkulla—Dr. Kelly's Tintara—Sound wine lore from the Antipodes—What missionaries should know about wine
pp. 151—158

CHAPTER XVIII.

Sparkling wines—Champagne, the great brands—Cheap sparkling wine from Saumur, Vouvray, Neuchâtel—the Styrian—Sparkling Hock, Moselle, and Tokay . . . . . pp. 159—161

CHAPTER XIX.

Fortified wine—Rogomme, Port, Sherry, Madeira, Marsala, Cape—General characters—High art and low art—Spirit added to Port; its effects—Characters of good Port—Tarragona, Roussillon—Characters of good Sherry, dry and sweet—Sickly and sad Sherry—Hypocrisy and tyranny of fashion—Madeira, Marsala, Malaga, &c. . . . . pp. 162—174

CHAPTER XX.

Derogatory estimates of wine—Laputan Philosophy—How to treat a cask of wine—Order of wine at dinner—Wine for the consumptive—Wine duties—Gratiarum actio post vinum
pp. 175—180

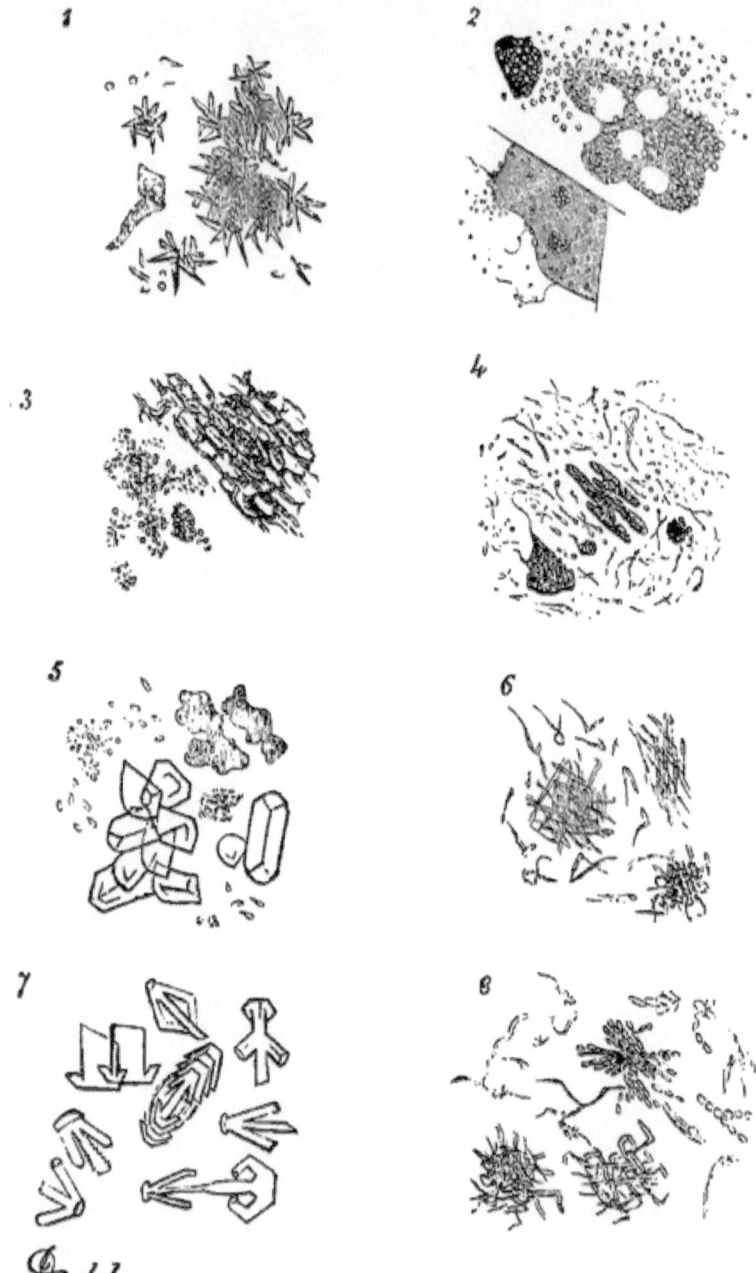

MICROSCOPIC DRAWINGS OF DEPOSITS IN BOTTLED WINE.
See page 60.

REPORT ON CHEAP WINES.

CHAPTER I.

Object of the book: vast scope of wine knowledge—Importance of it to the medical practitioner—Use of wine as a restorative—Wine often better than tea—Promotes sobriety and good morals—Empirical philosophy of wine—Knowledge of wine ought to be cultivated by all who buy and drink it.

N the following pages I propose to report upon the cheap wines which the public are now able to procure through Mr. Gladstone's remission of the wine duties, and the enlightened policy of the Emperor of the French. I am not going to write a treatise on wine in general,* for that would be far too ambitious a task and too voluminous. One moment's glance will show the multiplicity of branches which œnology or the

---

* See T. G. Shaw, on "Wine, the Vine, and the Cellar." 8vo, pp. 506. Longman, 1863. "The Vine and its Fruit, more especially in Relation to the Production of Wine," pp. 346. Longman, 1864. The works of Henderson and Redding, and Bence Jones's Translation of Mulder, are standard authorities on wine. The article "Wine," in the "Penny Cyclopædia," is generally known to be from the pen of my kind friend Dr. Robert Dickson, who may be pronounced the most advanced and practical œnologue in our profession. Years ago he demonstrated the necessity of *pure* wine in medicine and diet.

science of wine is divided into. There is Ampelography, or the knowledge of the innumerable varieties of Vine. There is Viticulture, or the science of the cultivation of the vine. There is the knowledge of the various processes involved in the Manufacture of wine, and the care of it till fit for use. There is the knowledge of the Chemical composition of wine, the various constituent parts into which the analytical chemist can divide it, and their proportions and qualities. Too closely allied with this is the science of the Wine-forger, the wretch who prostitutes his chemical knowledge to aid in the fabrication of sham wine, and the concoction of various fraudulent imitations. Then comes the skill of the Merchant, who knows what wine is wanted, and what the public should pay for it so as to remunerate him for his judgment in choosing wine, and the risks he runs. And I may say that there is no greater public benefactor than the honourable wine merchant who puts new kinds of wine within our reach. Then there comes the knowledge which the householder and consumer should possess, as to what kind of wine is wholesome and desirable to drink, where it may be had, and what it should cost. Lastly, there is the medical practitioner, who should know the virtues of wine as an article of diet for the healthy, and should prescribe what, when, and how much should be taken by the sick.

In the following pages I have endeavoured to supply information, not such as is needed by the cultivator or wine merchant, but simply by the public and the doctors, with just such a glimpse at wine matters as shall enable the courteous reader to join in any wine talk which may be going on in his presence.

It is most especially the interest of the medical practitioner who lives amongst a luxurious town population to have at his fingers' ends a knowledge of the character and properties of the chief wines of the day; and this is a branch of knowledge that the practitioner, even in the most primitive and rural districts, need not despise. We must take people as we find them. In large towns there are always a number of wealthy persons who think much of their dinner-table, and who would have a very mean opinion of their medical attendant if he showed himself not well acquainted with a gentlemanly style of living. I recollect, years ago, when beginning practice, how insignificant I felt, in comparison with Dr. ———, whom I met in consultation at the house of an important patient. This was not because I did not "know my profession," to use the common phrase, for Dr. ——— was one of a very old school; but his real *forte* was his knowledge of cookery and wines; and before that my light was soon put out. I took the lesson.

Certainly there is no reason why those wines which are abused by the gourmand in order to enable him to eat *too much*, may not be used by the medical practitioner to help patients who have a difficulty in eating *enough*.

A large proportion of the patients who come to most of us do so for some failure of nutrition. Be it in town or country, I will undertake to say that the number of invalids who require tonics is far greater than of those who need anything like depletion. The difficulty with delicate children is to get them to eat. There are the cases of "anæmia," "anorexia," and "debility," which figure by scores on the out-patient books of all the dispensaries that I have ever been acquainted with,—there are the agueish and the neuralgic affections of the poor, and the illnesses caused by hard work and exposure, by anxiety of mind, and those caused by child-bearing and protracted suckling amongst poor women. And in all of these cases some refresher to the appetite is needed. Besides, be an illness what it may, most practitioners finish off their patient with a "light tonic."

Now, what is a light tonic? A little dilute acid, a slight bitter, a small quantity of some aromatic, a little alcohol, and some fragrant ether. But this is just the "draught" that Nature has brewed ready to our hands in the fragrant and appetizing wines of France and Germany!

Surely if a patient has two shillings to spend on something that shall make him eat, he ought to be far more grateful to us if we provide him with a bottle of wine than if we give him a "mixture." I often used to prescribe a so-called Chablis at about 1s. 6d. a bottle, sold in Dean Street, Soho, and have found even poor dispensary patients satisfied with it. But, as I shall explain presently, patients must first of all be taught what *wine* is, and the right way to use it; and the difference between drinking pure *wine* as a refreshing beverage, and gulping down a *dram* of bad spirits disguised as wine,—such as the "People's own Sherry,"—just to create a feeling of warmth under their ribs. On these points I shall dilate presently. Meanwhile, in order not to be misunderstood, let me say that everything in Nature and Art has its use; and of course there are conditions which quinine and the more serious tonics, or which brandy can cope with, but which pure wine cannot.

But it is not merely in a medical point of view, but as a friend of sobriety and morals, and with a view of raising the status and culture of large classes of society, that I venture to advocate the larger use of *wine*—*i.e.*, pure wine—as a beverage. It might well take the place of a good deal of beer, tea, and spirits. There are large numbers of townspeople, and especially of women engaged in sedentary occupations, who cannot digest the beer which is so well suited to our out-door labouring population. The very tea which is so grateful to their languid, pasty, flabby tongues, from its astringent and sub-acid qualities, and which also comforts their miserable nerves, has this intense drawback—that when taken in excessive draughts, and without a due allowance of substantial food, it begets dyspepsia, and that worst form of it which impels the sufferer to seek a refuge in the gin bottle. Cheap wine would cut off the temptation to gin, and, with an equal bulk of water, would be found, in certain cases, a happy substitute for tea. I know a good deal of the better class of needlewomen and milliners' assistants, and speak from experience.

If instead of half the tea at the English middle class breakfast an earlier luncheon with a glass of light wine were given, it would greatly abridge the hours of half hunger, half dyspepsia, which many schoolgirls, shop girls, and other sedentary middle class women in towns suffer between their breakfast and early dinner.

For purposes of social exhilaration, amongst classes who are *not* out-door labourers, beer is too coarse. Man, as a social animal, requires something which he can sip as he sits and talks, and which pleases his palate whilst it gives some aliment to the stomach, and stimulates the flow of genial thoughts in the brain. No one who has ever made the experiment will fail to give the preference to wine over spirits,—

naked or disguised,—whether as gin or cheap incendiary sherry,—or can refuse to give a helping hand to any "movement" that will banish spirits to their proper place as medicines for the sick and aged, and not as beverages for the healthy. Civilized man must drink, will drink, and ought to drink; but it should be wine.

Let me say here, that the only real test for wine is the empirical one. It is impossible to dogmatize on it *à priori*; to say that such a wine, for instance, must be good in such and such cases, because it contains certain ingredients. The only questions we need ask are, not what is the chemical composition, but do you like it, and does it agree with you and do you no harm? The stomach is the real test-tube for wine; and if that quarrels with it, no chemical certificate and no analysis is worth a rush.

CHAPTER II.

What is wine?—Varieties of grapes: strong grapes and weak grapes—Anatomy of the fruit—Ingredients, saccharine and nitrogenous—Fermentation: process of, theories of, the chemical, the chemico-vital, or panspermatistic, and the archebiological—The new wine: the care it requires racking, sulphuring, and fining.

BY wine we mean juice of the grape, which has been fermented, and thereby has been partially converted into alcohol, and into various other stimulating matters. Some of its qualities depend on the grape, others on the fermentation and subsequent changes.

"*Il buon vino comincia nell' uva.*" Good wine is made from good grapes. The tree is known by its fruit. I am not going into the subject of ampelography or viticulture, yet it is a part of common wine-lore to know that the kind of vine is the most important element in determining the quality of the wine. Some vines there are which are hardy, grow rapidly, resist disease, and produce grapes in the utmost abundance; such is that which is called the crazy—*enrageat, folle blanche,* or *picpoule*; the extensive cultivation of which is lamented by M. le Dr. Jules Guyot.* This produces excellent brandy, but deplorable wine,—wine fit only to be distilled, or "burnt," as

---

* Sur la Viticulture du Sud-ouest de la France. Rapport à S. E. M. Rouher, Ministre de l'Agriculture, par le Dr. Jules Guyot. Paris, Imprimerie Impériale, 1863. p. 226.

our forefathers would have said. On the other hand, the vines which produce good wine are more delicate, require more care in cultivation, are more subject to disease, and what is worse, produce very much less wine per acre than the more plebeian varieties. Such are the *grenache, teret noir, pinot,* &c. Of the grand wines of Burgundy, the produce is seventeen to nineteen hectolitres, per hectare ;* of the inferior wines in the South they may get more than 100 per hectare.† This is, as I have said, a part of wine-lore of which the Englishman has little practical knowledge ; but I am obliged to point to it for this reason.

The object of my writing is to show the excellence of wine, or fermented grape juice, as a beverage. But unluckily, we are confronted with some of the scientific criminals, the felons of the laboratory, the forgers of the cellar and pirates of the stomach, who would fain persuade us that all the virtues of wine are due to one ingredient—alcohol ; and that therefore there is no harm in adding alcohol to wine, or making a sham wine by fermenting sugar and water with the husks of grapes from which the juice has been extracted. On the other hand, the practical philosopher sees in wine one whole and indivisible product of the grape, deriving virtues from the grape which cannot be got out of sugar and water. He therefore demands unmixed grape juice, and takes cognizance of the qualities which distinguish one grape from another. On this point let us hear the voice of M. Pasteur,

---

* Le Vin, par A. de Vergnette Lamotte, p. 14. Paris, 26, Rue Jacob. N.B. A hectolitre is about 22 gallons, and a hectare about 2½ acres.

† Guyot, Rapport, p. 205.

the great chemist and œnologue, and master of the doctrine of fermentation.*

"M. Bertholot," he says, "has expressed the opinion which, in my judgment, is correct; that the *vinosity*, or the power of wine, is not due solely to the alcoholic principle. Wine certainly contains one or more substances besides alcohol, which give it strength. I add, that these substances are not the produce of fermentation or vinification. They are all ready-formed in the grape, and it is easy to show that there are strong grapes and weak grapes, just as there are strong wines and weak wines."

He compares two kinds of grape, the *ploussard* and the *valet noir;* and shows that it is not mere quantity of acid and sugar that determine the acidity or sugariness of the taste of the juice; and that grapes may give a wine strong in alcohol and in acid, according to the test of the chemist, yet that this wine may taste flat and insipid, by comparison with wine from other grapes.

These are valuable words, and teach us the lesson that the apparent sourness of a wine on the palate is no true measure of the quantity of acid it really contains.

So much for the kind of vine. We may next linger for a moment on the grapes or berries.

These furnish a remarkable instance of the minute mechanism of natural objects, simple though they may seem to those who have never studied them deeply. The skin is a tough structure charged with wax, to preserve it from wet, and containing abundance of tannin, an astringent vegetable principle like

---

* Études sur le vin, ses maladies, &c., par M. L. Pasteur. Paris, Imprimerie Impériale, 1866.

that which is found in tea, oak-bark, and the like; and of colouring matter. The colouring matter is either deep blue or yellow. The blue colouring matter is found in the skins of what are called black grapes, and like other vegetable blues becomes red when mixed with acid, in the development of wine. The grape juice is not coloured, as any one may see who cuts across a black grape; it may be faintly pinkish, but it is not blue-black like new wine, nor red like old wine. Neither is it astringent; the colour and astringency of red wines are got by fermenting the skins of black grapes with the juice. The colouring matter is not soluble in water; ladies know too well that it will not easily wash out, when wine is spilled on a tablecloth; but it is easily dissolved in alcohol, and therefore is dissolved out during the development of alcohol in fermentation, along with the tannin. The deeper the colour of wine, the rougher it generally is as well, for the tannin and colour go together. Let the grape skins be macerated as they may, they retain abundance of colour to the last, and the men who shovel out the mark or pressed grape skins from the huge *cuves* or vats in which the wine has been fermented, are of the colour of indigo. No kind of artificial colour is so cheap and abundant as that of the skins. This may be a comfort to those who fear that French wines are artificially coloured with elder-berries or other pigments.

It is well known that the *mark*, or squeezed-out grape skins, when distilled, yields a considerable amount of brandy. Hereby we have an illustration of a fact well known to English ladies, that brandy-cherries are much stronger than cherry-brandy; that is to say, certain vegetable matters—such as cherries and grapes and other fruits, when soaked in a liquid containing alcohol, have the power of imbibing the alcohol into their substance, and thus contain within their interstices a liquid much stronger than that which they are soaked in.* Hence the expediency of not allowing the grape skins to macerate too long in the young wine.

The anatomy of the grape-berry is, on a small scale, the same in many respects as that which can be easily seen in the orange. Outside is the skin, which can be peeled off clean, as that of the orange can, and which if tasted is well known to be rough and astringent. It is thoroughly indigestible. The human gizzard will not dissolve grape skins. The flesh of the grape is a highly organized pulp, consisting of a mass of delicate cells or vesicles, filled with the precious juice. The juice of fruits, be it remarked, does not lie loosely in the interior as in a sponge, but as it were in a host of microscopic bladders; and in order that it may run out, the containing vesicles require to be thoroughly smashed up, just as is done in squeezing an orange or lemon, in which the containing vesicles are easily seen by the naked eye.

Inside the fleshy pulp are the pips or stones, some large and well developed, others dwarfed and abortive. These pips are of stony hardness, and abound in tannin. When we add that the stalks also abound in acid and tannin, we see the source of the roughness of the wines in which juice, skins and stalks are macerated and fermented together. We see the reason of the vigorous treading which was employed in former days, and still is in old countries to crush

---

* Guyot, Rapport, p. 229.

the grapes, because otherwise the juice does not run out of its containing cells; and why it is that the rollers employed to crush the grapes in Mr. Patrick Auld's vineyards in Australia are covered with cloth so that they may not crush the pips whose bitterness would be too much.

Having thus cleared the way, we may quote from the able chemists, Mulder, Maumené, Griffin, Guyot, and Vergnette Lamotte to tell us what ingredients they find in the grape juice.

First and most essential is *sugar*, the quantity of which goes on increasing in the grape as it ripens. The sugar is the *sine quâ non*, for without it there can be no alcohol. Intelligent and scientific wine-growers carefully watch the increasing quantity, day by day. The quantity of it in the juice is estimated by the specific gravity—*i.e.* its weight compared with that of an equal bulk of water, and there are besides more accurate, but more troublesome processes which we need not describe, merely adding that the sugar varies from 12 to nearly 30 per cent. of the weight of the grape juice or " must."

The second, and equally essential ingredient, is acid, especially the tartaric.

Thirdly, grape juice contains potass, an alkali derived from the soil, and found in the ashes of all land plants; some lime, soda, iron, phosphoric acid, and other inorganic principles common to all vegetables, and detected in the ash after they have been burned.

Fourthly, a number of substances, such as vegetable albumen, gum, odoriferous and flavouring matters, some common to all vegetable juices, others peculiarly characteristic of the grape, which owes to them its distinctive flavour.

So much for the grapes. The process of wine

making is begun by gathering the grapes, and crushing them; the juice is then put into *cuves* (Cuppa or Cupa) or tuns to ferment; for red wine, skins and often stalks, are all fermented together; but for white wine the juice is strained off and fermented by itself.

Now let us see what happens in fermentation.

The liquid becomes warm, it gives off carbonic acid gas in abundant bubbles, and when the fermentation is over it is found that the sugar has vanished, or nearly so, and that instead of it, the liquid from being sweet, is *vinous* and heady, and that if distilled, it gives off various substances more volatile than water, of which the chief, when collected, is known by the name of *spirit of wine*, or in its highest state of concentration *alcohol*.

Sugar is found in abundance in many vegetables, and there are many kinds of it, named from the plants in which each is respectively found—as cane sugar, grape sugar, and the like. The starch which abounds in vegetables (barley, potatoes, rice, &c., &c.), is also converted into sugar in germination or fermentation.

The main fact is, that when a vegetable juice is crushed out, or prepared artificially as in making a *wort* in brewing, if it contains sugar with the nitrogenous substances naturally found in vegetables, it naturally begins to ferment if exposed to air and sufficient heat. The first man who squeezed out grape juice, and tried to keep it a few days, was also the first man to make wine. The teetotal fanatics are not ashamed to say that God made the grape, but that the devil or man makes "cussed alcohol." Sane persons acknowledge that He who made the grape juice, gave it also the property of fermenting, unless hindered by art and care.

What happens in fermentation is very wonderful; it is, that sugar, a sweet, cloying, neutral-tasting stuff, is converted into almost half its weight of alcohol, a hot fiery heady volatile liquid, and half into the well-known gas called carbonic acid. This is rendered clear to chemical students by symbols. Sugar is composed of three elements, carbon, hydrogen, and oxygen $C_6H_6O_6$, in certain fixed proportions. Under the impulse of fermentation, part of the carbon and part of the oxygen C_2O_4 combine to form carbonic acid $2CO^2$; the remaining carbon and oxygen with the hydrogen form alcohol. Sugar $= C_6,O_6,H_6$. Carbonic acid $= C_2,O_4$. Alcohol $= C_4,O_2,H_6$. If I may quote the words of the most accurate, carefully composed, and scientifically exact work of Mr. Griffin,* "90 parts by weight of grape sugar produce 44 parts by weight of carbonic acid, which escapes as gas, and 46 parts by weight of absolute alcohol, which remains in the liquor." Hence, in theory, any quantity of grape sugar should yield a little less than half its weight of alcohol, and a little more than half of carbonic acid. It is here assumed that all the sugar is decomposed into alcohol, whereas in fact a small part remains; some is probably converted into other products besides alcohol, and part of the alcohol is lost. Anyhow the quantity of sugar is that on which the quantity of alcohol depends.

We must always, however, bear in mind that *pure* sugar *per se* dissolved in water does not ferment, but that, along with it, there must be supplied by nature or art some of the azotised albuminous or nitrogenous

---

* The Chemical Testing of Wines and Spirits, by John Joseph Griffin, F.C.S. London. Griffin, 22, Garrick-street, Covent-garden. 1866.

matter which is the essential constituent of all living things, animal or vegetable. The sugar in grapes is like the oil of the olive, or the fat of an animal, a most useful and beneficial product, but that which produces it is the nitrogenous living matter, *germinal* matter, bioplasm as Beale calls it, of which the actual substance of the plant consists, and which is the seat of its vital forces.

Given then, a solution of sugar, and the presence of albuminous matter as in grape juice, what is the force which sets it fermenting, and causes the marvellous change into alcohol and carbonic acid?

Two leading theories occupy the scientific world: one, which has descended from Willis and the great iatrochemists of the 17th century, and which is upholden in our times by the illustrious Liebig: this is the purely *chemical*:—the other, the chemico-vital, with which Pasteur's distinguished labours are associated, and according to which chemical change follows, on the development in the fermenting fluid of minute living plants.

According to Liebig's theory, a nitrogenous substance which is itself undergoing decomposition, is capable of effecting decomposition in any unstable substance with which it is in contact. The substance called *yeast* is a nitrogenous substance undergoing decomposition. The disturbance in its constituents during oxydation effects a disturbance in the sugar, which splits it up into alcohol and carbonic acid. "Alcohol and carbonic acid are produced from the elements of the sugar, and *ferment* (i.e. *yeast*) from the azotised constituents of the grape juice, which have been termed gluten or vegetable albumen." "Fermentation is excited in the juice of grapes by the access of air; but the process once commenced, continues until all the sugar is completely decomposed, quite independently of any further influence of the air."*

The *chemico-vital* theory which has been developed by M. Pasteur, is the more worthy of attention inasmuch as its scope has been extended, and it has been made to account for the decomposition of organic substances in general, and for those abnormal movements in living bodies which are known as fevers. The term *zymotic*—i.e., *fermentative*, is as we all know, applied to these diseases by the Registrar-General; and if Pasteur's theory be true, we have in the fermentation of grape juice, in the mode of limiting that fermentation, of preventing it from going to an injurious extent, and in the mode of preventing the various maladies and injurious changes to which wine is subject, the rationale of the mode of keeping our larders sweet and our bodies healthy.

M. Pasteur's doctrines, which are developed in those *Études sur le Vin* † which he was induced to undertake by the Emperor Napoleon III. (whom our volatile Gallic neighbours certainly cannot accuse of indifference to the physical well-being and commercial prosperity of their country), are briefly these:—It is assumed that the air, and the surface of almost every substance in the air, is laden with the germs of minute organisms, which become developed in organic substances which they gain access to; and the deve-

---

* Chemistry in its application to Agriculture and Physiology, by Justus Liebig, M.D., &c. &c. Edited by Lyon Playfair. Second Edition. London, Taylor and Walton, 1852.

† Études sur le vin, ses maladies, causes qui les provoquent, procédés nouveaux pour le conserver, et pour le vieillir, par M. L. Pasteur. Paris, à l'Imprimerie Impériale, 1866.

lopment of which in organic substances is attended with various changes of the nature of fermentation. Thus the grape juice, if squozen out under mercury, and with precautions to exclude the air, will remain unaltered; whilst if the smallest bubble of air be admitted, fermentation will begin. On the *chemical* theory, this is because of oxydation of nitrogenous matter, which communicates movement to the sugar, as we have before described. But according to the *chemico-vital* theory it is because the air introduces germs of the torula or yeast plant; this plant in growing forms itself out of the albuminous elements of the juice, and absorbs the sugar, which it decomposes, as above.

The details of fermentation are easily explicable in the chemico-vital theory; for instance, it is made slow by cold below 40° F., in which the yeast plant does not grow, and altogether stopped by heat above 120° F., which kills it. It is also checked or stopped by chemical agents—sulphur fumes, strong alcohol, creosote, and the like. It will not take place in liquids too dense or sugary; and it ceases when a certain quantity of alcohol has been produced.

We may say that it really seems established that the yeast plant is essential to the production of alcohol, but that in many other fermentative changes, mere chemical, and not chemico-vital agents suffice; whilst sometimes the presence of living organisms in decaying matter is a coincidence or adjuvant, rather than the prime cause.* We shall see further on how M. Pasteur's theory explains the diseases of wine.

---

* Pasteur's theory, which is a development of facts and doctrines set forth by Cagniard de la Tour, Turpin, and Mitscherlich, involves the doctrine of *panspermatism*, or the universal diffusion of the germs of microscopic animals and plants. The latest doc-

Well, to resume. The grape-juice fermented has become wine. Then it is drawn off into casks, where the fermentation may finish at leisure, so that all the sugar may be consumed, and all the nitrogenous matter capable of acting as ferment or as the pabulum thereof may be deposited.

The dangers which wine has to run the gauntlet of arise from unfermented sugar, undecomposed nitrogenous matter, and the germs of the microscopic plants which are found along with it; amongst these are the vinegar plant, and the filamentous growths that constitute the elements of the various wine diseases. Besides this, the excessive action of the oxygen of the air has to be guarded against.

The cares lavished on the wine during its first year, all have the purpose of meeting these calamities. The casks are kept filled; during the first few days they require to be filled up every day, then once a week, and always once a month. The loss of wine is great and constant; Mons. L. told me that let a cask be filled, and but rolled across the yard, there will be found space to fill up. New casks absorb some wine, the wood evaporates from its outer surface, and so it is calculated that a cask of 228 litres loses about one litre per month.

---

trine is that of *archebiosis*, set forth with consummate ability by Dr. Charlton Bastian. This affirms the origin of low organisms from decomposing nitrogenous matter; it takes life to be one out of many modes of chemical action capable of being propagated by contiguity; ascribes the origin of the yeast-plant not to the introduction of germs, but to spontaneous development in the nitrogenous constituents of the grape-juice; and explains the absence of fermentation in liquids from which air is excluded, by the simultaneous exclusion of the dead nitrogenous particles floating in the air.

Then the wine in its first year is submitted to two, three, or four *rackings* or *soutirages*—that is, it is drawn off from the lees into another cask. This is done when the weather is cold and the barometer high, because then the residue of nitrogenous matter has less tendency to rise and mix with the liquid.

Another conservative operation to which the wine is subjected is the *méchage*—i.e., burning a brimstone match in the empty cask to which the wine is to be transferred. This is well known to check fermentation; on the chemical theory it acts by abstracting oxygen; on the chemico-vital theory it acts by killing the germs of the yeast plant and vinegar plant, and others which cause wine disease. On the theory of archebiosis, it makes the nitrogenous particles stable, and incapable of spontaneously developing life.

Another operation is *fining*; which consists in adding to the wine some matter that shall curdle and contract, and entangle and carry with it to the bottom all floating particles which make the wine thick; and of course, as it clears the wine of particles of ferment, organized or unorganized, so it tends to make it not only bright to the eye, but better able to keep.

It will be seen that filtration through the finest paper is an effective mode of clearing wine of minute germs, and how intimate the connexion of *condition*, or brilliancy, is with soundness and wholesomeness. The fine wines of Burgundy are sometimes *frozen* in order to make them deposit excess of colouring and nitrogenous matters, and deprive them of the germs that cause secondary fermentation. Whilst frozen, the concentrated and purified wine is drawn off from the net-work of watery icicles and impurities which remain in the cask.

CHAPTER III.

Wine as a subject of chemical analysis—Its proximate principles —Alcohol and its varieties: aldehyde, ethers, acids, earthy salts, glycerine—Abundant sources of alcohol—"Strong drink" of the English Bible—Percentages of ingredients.

NOW for a few words on the chemical ingredients of wine. They who desire to know this matter fully, and to perform analyses for themselves must study Mr. Griffin's admirable book, clear as a bell, short as wit itself, with not a word out of place, and full as a new-laid egg. Mine is a popular book, and I only care to give such details as shall enable the consumer to know what wine ought to be, and to induce him to drink it.

Chemical analysis detects and separates the various ingredients in bodies submitted to it. Those ingredients into which the body examined most readily splits, are called *proximate;* and these can be more and more split up till we come to *ultimate* elements which defy further analysis. Thus a plum pudding is a compound body. Its proximate ingredients are flour, suet, raisins, &c. The flour is a compound of starch, gluten, &c., and these may be analysed and found to consist of Carbon, Hydrogen, Oxygen, &c., which are called ultimate elements because they cannot be analysed any further.

The simplest analysis of wine is effected by heating it, for then there will soon be proof that there is in it a something which flies off in the first steam, and which can be entirely driven off by long boiling. If the steam be conveyed through a proper apparatus, and cooled, it will condense into a liquid which may be collected, and is known as spirit of wine.

Spirit of wine so obtained, is more or less contaminated with other volatile matters, and notably contains a considerable quantity of water. When purified by fresh distillation and various chemical processes, a liquid called *absolute alcohol* is obtained, very volatile, fiery to the taste, very light, having a specific gravity of ·794, when an equal bulk of water would weigh 1·000. Absolute alcohol is to most of us an ideal substance, a chemical curiosity, too expensive and volatile for common use. When it is diluted with about 16 per cent. of water and is of sp. gr. 838, it is called rectified spirit, and this is the strength in which it is found in strong Eau de Cologne, Tincture of Myrrh, Essence of Camphor, and the like. When a spirit contains half its weight of water, and is of the specific gravity of ·920, it is called *proof spirit*. This is about the strength of ordinary gin and brandy, and is taken as the standard. When we speak of wine having such and such a strength,—say that claret contains 17 per cent. of proof spirit,—we mean that seventeen parts out of 100 of its bulk or volume, consist of a spirit of this strength. When a liquid contains more absolute alcohol than proof spirit does, it is said to be so many degrees over proof; if of less strength, so many degrees under.

Pure spirit and water has very little flavour and no body; it is a thin unsubstantial drink, but gives a sensation of heat in the throat. For use, spirit is always flavoured and often coloured, as in the form of whisky, brandy, gin, Hollands, and the like.

Any vegetable sugary, or starchy matter may be made to ferment, and in different parts of the world fermented drinks are made from milk, apples, pears, honey, barley, wheat, maize, rice, plums, cherries, and the juice of the palm tree. These are enumerated by St. Jerome as common drinks, in a well-known and most valuable letter to a young priest. In the Bible we read of "wine" and *"shekar,"* or *"sikera,"* or *"sicera."* Wine was the juice of the grape; *sikera,* (from which our word cyder is said to be derived) included every other kind of fermented drink—beer, cyder, mead, palm-wine, or toddy, and the like. In the English authorized version of the Bible, the word *sikera* is translated "strong drink," which is unfortunate, as some people believe it to mean distilled spirits; but these were unknown to the ancients, and were invented by the Arabians; *alcohol* is an Arabic word. The advice which St. Jerome gives to the young priest ought to be printed in letters of gold. He tells him "to take care never to smell of wine." A priest should have the character of abstinence, and should never either set an example of aught save the strictest temperance, or follow it. Moreover, a young man, full of life and blood and spirits, does not want wine, but should reserve it for occasional legitimate festivity, or for the time when age, sickness, or weakness may demand it.

I who write in the praise of wine must also caution the young and healthy, that it is scarcely needful for them as a daily drink, if they can take nourishment enough without it.

But to return—Spirit of wine, or alcohol, is but one substance of a class. An alcohol is a compound of some multiple of C_2H_2 with H_2O_2. Common alcohol, called also *ethylic*, is C_4H_4 with $H_2O_2 = C_4H_6O_2$. But there are other alcohols containing other multiples of C_2H_2, as the amylic or potato alcohol, the methylic or wood-naphtha; and some of them are apt to be produced in small quantities in ordinary fermentation, and give to raw and unrectified spirits peculiar characters, and are for the most part very heady and intoxicating. These are commonly called *feints, fusel oil, &c.*

The next element in wine is its acids, of which more hereafter. Whereas acetic acid is produced by the entire oxydation of alcohol, there is a substance called *aldehyde*, which is formed by such a partial oxydation as shall take away some hydrogen, and produce a body intermediate between alcohol and acetic acid. Its presence in wine, says M. Maumené, is of real importance, both as regards flavour and taste.

In the next place, when any alcohol finds itself in the presence of any acid, part of it is convertible into a new compound called an *Ether;* and as the ether of every alcohol is able to form a compound with any acid, and as there are half-a-dozen alcohols and a dozen acids, so there is the probability of almost any number of ethers in wine. The most familiar example of a compound ether is the *nitrous ether* or *sweet spirit of nitre* sold at the chemist's, a volatile, pungent, fragrant, fruity liquid, which is largely used by people who have "taken cold," and by whom it is often called "nitre." The most distinctive of these wine ethers is called *œnanthic ether*, which is easily procured from wine lees by distillation. It is a liquid of exceedingly heavy, repulsive, disagreeable odour. It must be remembered that many substances require to be presented in very small quantity in order that their beneficial qualities may come out. Thus the otto of roses and oil of peppermint, of orange, lavender, and the like, might scarcely be recognised, save as strong and disagreeable substances, when presented to the nose in large quantity undiluted. How small a quantity of garlic will give a rich, full, savoury fragrance to a leg of mutton! The same in excess would be pronounced detestable by any one who had not got over his Anglican prejudices. *No qualities are absolute*, but must, amongst other conditions, depend on quantity. It will, therefore, not be surprising that the quantity of these subtile ethers in wine is extremely low. When Maumené states that it cannot be more than one part in a thousand, but possibly is as low as one in 200,000, he reminds one of the philosophers who attempt to calculate the age of the sun.

Any one may watch the phenomena of wine-making who will take the trouble to crush a dozen grapes, put them into a wine-glass, and cover it with an inverted tumbler to keep out flies and dust. In warm weather, bubbles of gas are soon seen to form; a sediment sinks to the bottom of the grape juice; the grape skins rise, and form a *chapeau* on the top, and the petty *cuve* soon yields an unmistakeable smell of wine: a thing *toto cœlo* different from the mere smell of grapes; and it must be confessed that a strong wine smell is very disagreeable; for example, when a cask of wine is tapped, the house is not pleasant. No more is it if filled with the vapours of boiled beef and cabbage. The moral is, that everything has its place and is subordinate to due conditions, one of which is quantity.

Next to the ethers, formed by the action of acids

upon alcohols, we must reckon the odoriferous matters originally contained in the grape.

When we have enumerated the water, alcohols, ethers, acids, sugar, colouring matter and tannin, we have given those ingredients in wine that are most palpable; but the number is not by any means exhausted. Oil and wax are derived from the skins and husks; glycerine, the well-known sweet emollient, was announced by Pasteur, in 1860, as a product of fermentation both in wine and beer. He says that he has detected 6, 7, and 8 grammes per litre—(a *gramme* is a little more than 15 grains, and a litre = 1000 grammes, or about one pint and three quarters, so that a litre of wine would contain about 85 grains, or a very large teaspoonful of glycerine)—whilst, he says, old wines that have undergone a good deal of evaporation by having been long kept in cask may contain so much as 10 or 12 grammes per litre. Succinic acid is another substance formed in wines. Besides the matters found in grapes especially, or developed by fermentation of grape juice, there are those which are common to all vegetables, which we have spoken of when speaking of grape juice. Above all these is potass, which has the admirable property, when combined with two equivalents of tartaric acid, of forming a salt, called bitartrate of potass, or cream of tartar (in its rough state it is called *argol*). Now this substance is less soluble in alcohol than in water, and therefore, as alcohol is developed in wine, so the cream of tartar is slowly deposited, and continues to be so for some time. It forms those brilliant crystals which are seen on allowing the cork to dry, after opening a bottle of any wine which has been bottled young. The neutral organic substances, including sugar and tannin, and various unknown substances lumped together under the term *extractive*, form what is called the body of wine.

The quantity of solid matter which any wine holds in solution is governed chiefly by the quantity of sugar, which in sweet wines may be 20 per cent. or more; but in old fine wine, dry and free from sugar, the quantity may be very small—not more than 2 or 3 per cent. or none at all.

When the solid residue of wine is burned with exposure to the air, there is left a minute quantity of *ash*, consisting chiefly of potass, the alkali found in all vegetables, with phosphoric acid, lime, iron, manganese, and other earthy bodies in minute proportions.

M. Maumené gives as the average composition of a pure wine, per thousand parts by weight, 891 parts of water (*i.e.*—89·1 per cent.), 79 (or nearly 8 per cent.) of absolute alcohol, and 30 (or 3 per cent.) of particles which constitute flavour, taste, and odour, including extractive and salts. If we reckon up alcohols, ethers, volatile oils, vegetable extractive, albuminous matters, acids, and alkaline and earthy matters, the industry of chemists has made a list of more than one hundred separate substances found in wine.

The functions of the astringent matter deserve one word. Everybody knows the preservative virtues of oak-bark: how skins steeped in it become *leather*, which is almost imperishable, and how human remains have been preserved when coffined in a hollowed oaken trunk: what virtue vegetable astringents have in hardening and condensing the tissues of the living body, so that the blood does not leak out of the small blood-vessels nor the vessels break under the pressure of the blood, hence their use in hemorrhage, profuse perspirations, and the like. Most tonics are astringents, as, for example, the Peruvian bark, which cures ague; but other astringents do good in ague, and a powder of alum is very popular in some of the fen districts. The astringent matter of red wine is therefore an important ingredient as regards the effect of the wine on the drinker; but still more, it is good for the wine itself; —it keeps the wine elements together and hinders certain changes, and acts as a cure of certain degenerations of wine. Hence it is often added, but the most refined plan of doing this is to use an infusion of grape pips; the tannin produced from oak-bark or from nut-galls has a certain flavour and bitterness which are easily detected by any one who has tasted them. The tannin is the chief agent in the formation of the *crust* or deposit of which wines " despoil themselves," and which consists of colouring, extractive, and astringent matters in an insoluble state from slow oxydation. The oxygen of the air, acting in a slow limited way, is, as M. Pasteur says, that which matures wine. As time goes on the colour of the wine becomes tawny, its sweetness, if any, diminishes, and it acquires the taste and flavour of *old* wine.

CHAPTER IV.

Wine and its effect on the nerves of taste—Taste, and smell, and touch—Specimens of wine tasted—A light white Bordeaux wine; a light red St. Estèphe; a fortified red wine from L'Hérault; Château Lafite, 1851; Château Yquem, 1858; St. Elie—Varieties of flavour—Rules for tasting wine—Alcohol affects the nerves of touch—Vocabulary of wine terms—Difficulty of describing sensations.

HAVING waded with more or less patience through the very hasty sketch of wine making and wine composition, which I have attempted in the foregoing pages, the courteous reader may now come to the properties of wine, as shown by its action on the animal economy.

The organs of taste and smell stand as sentinels to watch the approaches to the stomach, and to warn us whether our food and drink are fit to be admitted or not. There are some articles respecting which these organs are not entirely to be relied upon; but certainly as regards wine, the effects of wine on the palate are known with exactitude, and the palate is able to distinguish wines which are wholesome from those that are not.

Let us observe that *touch* is common to all parts of the body in greater or less degree, but is especially acute in the finger-tips, lips, and tongue. This takes cognizance of certain qualities, such as hot and cold, rough and smooth, hard and soft, and the like. *Taste* is a more delicate sense, and distinguishes properties such as sweet, sour, bitter, and salt, together with a thousand other varieties which have no name, though we well know them when presented to us.

There is a third sense which recognises odours, and upon which they particularly operate, of course I mean the nose. Now everything that is tasted must affect the sense of touch, and the union of both touch and taste may be essential to perfect enjoyment; thus, the crispness or flabbiness of a biscuit may make a great difference. Just so the union of smell with taste is essential for the enjoyment of wine. And here let us say, that everything that is smelled can be tasted, though not everything that is tasted can be smelled. The body of wine affects both senses.

Now let me advise any beginner to make this little experiment beforehand. Take on the end of the forefinger a few petty crystals of salt. Touch with it the inside of the upper and lower lips; there will be only the sensation of touch. Touch the palate behind the upper teeth,—no other sensation. Carry the finger far back to the soft palate quite at the back of the roof of the mouth; then there will be a clear and distinctly localized salt *taste*. It is very distinct on the uvula; not on the tonsils. Touch the tip, edges, and back of the upper surface of the tongue, and the salt taste will also be perceived; and in making an effort to swallow, it will be noticed that the taste is most intense when the back part of the tongue and the soft palate over it come into contact during the act of swallowing.

Armed with this physiological information, let me go down into the inmost compartment of my modest *cave*, where my eye falls on a bottle whose label is *Langoiran blanc*, 1862; *Caves de la Gironde, sous la direction de G. Fauché fils, et C. Brisac, Propriétaires à*

Bordeaux. This is one of a hierarchical gradation of samples of Bordeaux wines which MM. Fauché fils and C. Brisac were good enough to send me in 1866, on the publication of the former edition of this book, in order to contribute to the study of œnology from good specimens. Let us carefully draw the long soft cork, and decant the wine from a small quantity of flocculent sediment visible in the bottle, and then pour a large glass half full and drink a huge sip of the pale primrose-coloured wine. The effect is agreeable. The tongue and the soft palate are conscious of a slightly acid, yet soft liquor, eminently refreshing, producing a pleasurable sensation all over the inside of the cheeks and the hard palate and throat; and if the mouth be kept shut, as it should be, there is at every breath breathed out, a sense of perfume wafted upwards and detected by the throat and nose. The taste and perfume are those of fresh green grapes. There is in the taste no heat, no bitterness, no astringency, no unpleasant sourness; there is a sense of delight and satisfaction and well-being, which is diffused over the whole body. The price in 1866 was 1fr. 75c. at Bordeaux. This is a good specimen of a good ordinary wholesome white wine. The deposit submitted to microscopic examination, displays a few crystals, and a few bodies that look like torula, and a small quantity of filamentous cobweb stuff. This represents the residue of the albuminous part of the grape juice, which had not quite separated when the wine was bottled. Be it observed, that though the wine is ten years old, it is as *fresh* as if its grapes had been gathered yesterday.

It will be noticed that this was a white wine—a simpler thing than red wine. So let us pitch upon a bottle of light claret from Bordeaux ; a *vin rouge St. Estèphe*—one of the lightest, thinnest wines I ever met with. Of a very pale ruby colour and distinct vinous smell, the faintest astringency, and just marked acidity, we have a type of a fine ordinary Bordeaux wine. There is a taste *sui generis*, which we know as the taste of wine, and there is an entire absence of anything positively unpleasant, whether sweet, sour, or rough. It is marvellously thin and destitute of body, but is as sound as possible. The flavour is slight but very good of its sort.

Next let us have a thorough contrast ; a bottle of red wine from the South of France—the black red colour and the powerful odour would distinguish it at a glance from the somewhat *etiolate* specimen we tasted last. The moment we taste it we are conscious of *heat*, a heat that diffuses itself over lips, cheeks, tongue, palate, throat, and that can be felt down every inch of one's backbone as it travels to the stomach. This is a sign that the wine has been *viné*— *i.e.*, mixed with spirits, and pretty handsomely too. We need not trouble to analyse it ; we may be sure that it contains more than twenty-six degrees of proof spirit. Under this heat we discover a powerful vinosity, great body, great astringency, some flavour, and little or no sweetness. This is a South of France wine, from the cellar of an accomplished and patriotic proprietor ; but it is a wine to sip, not to drink at a draught.

Now for another contrast. Here is a bottle from the *Caves de la Gironde*, entitled *Château Lafite*, 1851. The price at Bordeaux in 1865 was ten francs per bottle. The colour is brilliant dark ruby, like the last. But this is cool, it is absolutely devoid of heat. The impression on the palate is of great *body*—*i.e.*, that the wine holds in solution a large quantity of matter capable of being tested by the nerves of taste. Acidity is present but not felt—*i.e.*, if one thinks of it, one confesses that the wine is acidulous and not neutral, but there is no prominent acidity. There is no sweetness, some astringency without roughness, and a powerful fragrance after deglutition; the fragrance seeming ultimately connected with the *body* which is felt by the palate.

Let us take another specimen. A patient with lung malady tells me that he finds that the wine which agrees with him best, is Château Yquem which he imported from the Château ten years ago, and asks me to taste it. The well-fitting cork bears the brand "*Château Yquem*, 1858." The wine is of a pale straw colour; it flows as it were thick; when tasted, so far from the palate seeming to appreciate all the taste there is, as it does with a thin wine, its body seems inexhaustible; its flavour is fragrance of a thousand colours. Is it sweet? No, the soft nectarous taste is not mere sweetness, though decidedly fruity; there is just a perceptible after-taste of bitterness which seems to say that the wine has attained its maximum of excellence. Though soft, it is abundantly alcoholic.

Let us try one specimen more. Here is a bottle of St. Elie, a white Greek wine, which has been in my cellar awaiting its destiny since 1864. The wine bright brilliant dark amber colour; when sipped, one is struck with the feeling of body the abundance of vinosity; all the organs are gratified, and during the act of deglutition, there is the feeling of power coming into one; yet how different the taste and flavour of this wine, from that of the Langoiran with which we began; all the difference between a pretty young girl and a wise old man. The *substratum*, the *fons et origo* of the St. Elie is a honey-flavoured grape, and this is still detectible; the wine at first was harsh and sour, now it is soft. I see that Mr. Griffin in his before-quoted work, describes the "*St. Elie.* A dry white wine; taste slightly acid like that of Rhine wine, but with a flavour resembling that of Madeira." So it is with sensations. We cannot at present coin new words to express sensations; thus it is that we are forced to call some wine *grapy*, and to compare others to a wine already known, such as Madeira.

Wine being grape-juice fermented, should taste of grapes. Some wine has clearly come from grapes greenish, and not over-ripe, as Moselle and Rhine wines, Langoiran, Dioszeger, Œdenburg, &c. Some wine tastes of grapes fully ripe, mature, yet not luscious, as Ofner; some of grapes dead ripe and luscious, as Tokay and Château Yquem; many of grapes with musky perfume, as Muscat, &c.; some of grapes dried, as Cyprus and Visanto; some of grape-juice concentrated by boiling, as Como, brown Sherry, Tent, &c.; some of withered grapes, as certain Sherries; some of no grapes at all.

Let me now give a few rules and observations on that which ought to be looked for in tasting wine.

In the first place, in drinking a good large sip of the wine, does it *primâ facie* strike us as being *one* liquid, or a compound of many? Wine should have an absolute *unity*, it should taste as one whole. True, we may distinguish various properties on reflection, but they should be as parts of a whole, and not as independent units mixed together. But bad wine resembles what schoolboys of the last generation used to get when ill—a "black dose." Here a something sweet meets one part of our gustatory organs, there something sour, there something fruity, or bitter, or hot, or harsh, just as if half-a-dozen ill-blended liquids came out of one bottle, with perhaps a perfume atop, that seems to smell of a hairdresser's shop.

2. Wine should have a certain generosity of taste; unless spirit could be extracted by distillation it would not be wine; but there should be no smell nor taste of added and ill-combined spirits, nor that heat about the throat which they cause.

I must take the liberty of enlarging on this point. Spirits have their uses, but the very purport of my book is to show that pure good wine is the drink for healthy people in the English climate, that it has all the virtues and none of the vices of spirits, that it is not liable to the risk of excess, and that spirits, whether *neat* or in the shape of fortified wines or liqueurs, should be kept out of the way of every one who is not driven to them by age, infirmity, or disease.

Let me descant on the kind of sensation which alcohol gives. It is a simple warmth; it does not affect the *taste, quà* taste—*i.e.*, the power of distinguishing the sapidity of food and drink; but it burns—it burns every part alike by virtue of its physical qualities, and the sensation produced by it is not peculiar to the sense of taste, but common to the whole body. It burns lips, gums, hard palate and throat, just as it does the parts of the palate specially endowed with *taste*.

Some years ago the great experimental physiologist Magendie (the hellish Magendie, as Beckford called him) made experiments by vivisection to determine the functions of various nerves. The nerves of smell were cut through within the cavity of the skull of a dog, then strong ammonia vapour was applied to the membrane of the nose; the miserable, mutilated animal gave signs of sensation; *ergo*, argued Magendie, the creature smells, and smelling does not depend, at least exclusively, on the nerve designated as the olfactory, or nerve of smell. Magendie, by the way, said of the animals he used in these accursed experiments, "*ils ne s'amusent pas ici!*"

But a more acute physiologist showed that Magendie's experiment was ill-devised, and his conclusions wrong, for there are general sensations and special ones. General sensibility, or touch, is an endowment of most parts of the body, and tells us whether what approaches us is hot or cold, hard or soft, oily and smooth, or harsh and burning ; thus, whether ammonia be applied to nose, eyes, mouth, or back of hand, we shall be conscious of considerable burning force. And so in the experiment, the nose, though utterly unable to appreciate an odour, yet was painfully affected by the caustic vapour, as it would have been by a hot iron or any other appeal to the sense of touch.

Just so spirit of wine uncombined acts as a stimulant alike on the eye, the nose, and the mouth; and the physical sensation which it creates in the mouth and throat may be agreeable to some people, but is a most dangerous and unfruitful kind of pleasure; it only admits of the degrees of hot, hotter, and hottest, or very hot; it gives no scope for the intellect or the imagination, as does the study of the divine taste and flavour of wine. It is stimulation pure and simple, a something that sets the current of blood flowing, just as mustard does (some people, by the way, like the stimulant effect of a mustard poultice on the skin); and the effect of spirits on the palate is like it.

3. Wine, like all drinks used by healthy grown-up men, is slightly sour—not even excepting water, if it contain a palatable quantity of carbonic acid and dissolved chalk. All soft, neutral, or alkaline drinks are, like milk, adapted for infants, or, like Seltzer water, for invalids, or for people past their grand climacteric, or for the gouty. But all the drinks of healthy men and women are sour, such as tea, coffee, ale, beer, cider, mum, mead, perry, every kind of fermented drink known to the law, including wine, of course, and all the fruits which bountiful Nature gives us. So, too, are meat and vegetables in a lesser degree; flesh, fish (less so), bread, the horse-radish, the potato, the carrot, and the like. *Nature abhors alkalinity.* A certain amount of sourness belongs to all wines, and we have it *naked* in the well-fermented wines of France and Germany, and disguised in the imperfectly-fermented and sweetened and fortified wines of Spain, Portugal, the Cape, &c. There are other legitimate *sweet wines*, as Tokay, which, though they contain an exuberance of sugar, are well fermented and not brandied. The degree in which the natural sourness of wine affects us depends much on the state of the palate. Divine instinct teaches most men (who have not coddled themselves after the teaching of dyspeptic physicians and writers on diet) that something sour is good with fish and other gelatinous things, and with what is fat and high flavoured—not good with sweets or fruit. But an excessive sourness, depending on acetous degeneration or *prick*, is bad, especially if combined, as it is in some of the worst clarets I have tasted, with great alcoholic strength and taste of added spirits. Practice soon detects this.

4. Sweetness is a characteristic of many good wines, as I have just said; but the Bordeaux and Rhine wines are as nearly as possible *dry—i.e.*, not sweet.

5. Whether sweet, or dry, or acidulous, we look in wines for a certain *stability*—a *clean, round, firm, perfect* taste—and for the absence of what indicates change or fermentescibility. There is a certain mawkish, sickly sweetness (it is most like that of the sugar which is apt to *candy—i.e.*, to crystallize out of over-kept preserves, such as currant jelly), which, once tasted, can never be forgotten, and which indicates want of firmness. This may be combined with the *moustille*, the slight *pricking* sensation caused by the incipient disengagement of carbonic acid.

6. Roughness or astringency is a most important property, and belongs to most red wines. In moderate degree it is relished, as sourness is, by a healthy manly palate, just as the cold souse is welcome to the skin. In excess it leaves a permanent harshness on the tongue. The *genus* dandy always hates anything rough, but I believe a fair amount of it is anything but unpalatable to the tongue of a hardworked man. The careful old man in Terence's " Heauton-timorumenos " complains of a luxurious young lady in his house :—

> " Pytissando modo mihi
> Quid vini absumsit ! sic : ' *hoc*,' dicens, ' *asperum*,'
> ' *Pater, est ; aliud lenius, sodes, vide !*'
> Relevi dolia omnia, omnes serias."

" What a lot of wine she wasted in tasting it ; saying, 'This, papa, is rough ; see if you can't find some softer,' so I had to ransack and tap every bin in my cellar." Just so the young men at the clubs now ask for *some that is softer*. Civilized man is the same in essentials at all times and places.

7. In the next place we look for *body*. This is not strength, though the fullest bodied wines, as a rule, are the strongest. Spirit and water has no body. Body is the impression produced by the totality of the soluble constituents of wine—the extractive, that which gives *taste* to the tongue, and which, as wine grows older, is deposited along with the cream of tartar, forming the *crust*.

The *body* of wine resembles what we should call the strength of *tea* or any similar infusion. Suppose we taste an infusion of tea, or coffee, or hops, or any other powerfully-tasting substance, the tongue will soon tell us whether there be much or little dissolved in it. So it does with wine.

8. Next come the odoriferous principles which constitute the soul and glory of wine, and its distinction from other liquors. Some flavours are derived direct from the grape; a pleasant reminiscence of the fresh fruit. Other flavours are derived from fermentation; and of these the essential one is that by which we know wine to be wine. But, in addition to this, are those flavours perceived in the throat and nose, which are developed as the wine grows older, and especially in bottle. I need scarcely say that these last are not to be looked for in young wine; in which fulness and colour, if combined with purity and firmness, are to be chiefly looked for, and which, if kept a year or two, will improve vastly and get flavour. I am speaking throughout of moderate-priced wines fit for families. *Bouquet* is that quality of wine which salutes the nose, and very high bouquet in cheap wine is suggestive of perfume added by adulteration. *Flavour* is that part of the aromatic constituent which gratifies the throat. Here let us quote from Maumené,[*] and recall to the

---

[*] Travail des Vins. Paris, 1858, p. 522.

reader that the word *bouquet* signifies a nosegay, or bunch of flowers, and that, although it may be used to signify a bunch of one kind of thing, as a *bouquet de persil*, or bunch of parsley; still that, properly speaking, it means a bunch of divers sorts of flowers; and, therefore, to speak of the bouquet of wine as though it were one thing is " *une pensée tout à fait inexacte. L'arome du vin est un mélange très complex.*" The various alcohols, aldehydes, ethers, and oils combine to the result, and there are in many wines the chemical elements which can be combined by art into imitations of well-known flavours,—notably the butyric ether smells of pineapples. The finest bouquet I ever was conscious of was in a bottle of *Rauenthaler*, 1846, sent me by J. D. F. The grape, the peach, the strawberry, and the pineapple were all brought to mind.

9. The wine must *satisfy*. A man must feel that he has taken something which consoles and sustains. Some liquids, as cider and some thin wines, leave rather a craving, empty, hungry feeling after them. To wine, above all other kinds of food, we may apply the phrase of the Hebrew poet—" It satisfies the hungry soul, and fills the empty soul with goodness." Amongst the senses we ought to reckon the consciousness of hunger, emptiness, and exhaustion; and of the opposite, comfort, fulness, and well-being. Dr. Edward Smith, in the course of observations on the effects of food, notices how, when a man is intensely hungry, an innutritious food, such as arrowroot, which does not supply the real material wanted, is tasted with disgust or nausea, whilst a bit of bread, the true staff of life, the moment it is bitten and tasted, seems to create an instant feeling of comfort and satisfaction that is felt down to the ends of the fingers. There is a sense in the whole body, that the

exhaustion and emptiness are in way of being relieved. So it is with wine. Before even it has reached the soft palate, so soon as it is felt by the sense of touch upon the lips, there seems an electric telegram diffused over the remotest twig of every nerve, that a something has come to sustain a feeble heart, fill empty veins, feed and vivify an exhausted brain, and make the work of life go smoothly.

I have hitherto described in my own words the sensations which wine produces on my own uninstructed palate. But I ought not to leave my readers in ignorance of the large and exquisite set of phrases which the French have devised for the description of the qualities of wine. He that would study the subject as it deserves should get Monsieur de Vergnette-Lamotte's book *Le Vin*. Ladies and children may study M. Maurial's cheap little book, *L'Art de boire, connaître et acheter le Vin, &c.**

The *taste*, *saveur*, or *goût* signifies the effect on the nerves of taste in the palate. The *arome* is the effect on the throat; the *bouquet* on the nose. A *vineux* wine is a wine alcoholically strong. The *sève* is the physiognomy and whole character of wine, its *manière d'être*. "Whilst," says M. Maurial, "the *bouquet* gratifies the nose, the *sève* gratifies the mouth and stomach, and survives the deglutition of the wine. It is a *saveur spiritueuse et embaumée*, which is a quality of great wines, without any of that alcoholic odour which ordinary wines give."

A wine is said to be *corsé*, to have *corps* (body, stoutness, or substance) when its constituent parts intimately combined seem to make one complete

---

* *Le Vin* is published in Paris, 26, Rue Jacob, fr. 3.50; Maurial, at 112, Rue Richelieu, 1865, fr. 1.20.

whole. A *vin corsé* resists time, and deposits crust without getting thin. *Dureté* (hardness?) in wine means a deficiency of that unctuosity which masks the roughness of various saline and astringent particles—tannin and tartar. Great *dureté* is often combined with great solidity. When *corsé* and *dure* wines improve by age they acquire a *rondeur*. Wines of *dureté* with ordinary taste are called *rude*. *Apreté* is produced by excess of tannin as in the common wines of Bordeaux; *acerbité* recalls the taste of unripe grapes; wine is *acide* when there is excess of tartaric or malic acid, and *aigre* when there is the taste of vinegar. *Verdeur* is the sour taste of young wines made from unripe grapes. *Finesse* is a delicacy of taste, an absence of all that shocks the palate. *Velouté, soyeux, moëlleux,* are terms to indicate the soft, silky, velvety, marrowy character of full-bodied wines that are neither sweet nor dry; they indicate absolute homogeneity of the constituent particles.

A wine is said to be *vif* when it seems at once to penetrate the recesses of the gustatory apparatus; the *vivacité* is the opposite of *lourd* and *huileux*. A wine is said to be *plat* when there is an absence of alcoholic taste, and *faible* or *aqueux* when there is also a watery, insipid taste. A *dry* wine (*sec*) is one that has lost its unctuosity and sweetness: the terms *rapé, decharné, depouillé par l'âge,* are also used to indicate a loss of the more pleasing and soft qualities. *Franchise de goût* signifies the absence of everything earthy;—of the *goût de terroir,* that indescribable something which Cape wines used to taste of,—and the absence of all other taste indicating decay or disease in the grape or the wine, or mouldiness in the cask. The *moustille* or *goût de fermentation* is the pricking sensation occasioned by a small quantity of carbonic acid; it shows instability in the wine. The *goût de cuit* is, says De Vergnette-Lamotte, that artificial taste which characterizes the sweetened wines of Spain. It comes from the conversion of grapes into what we call *raisins*, or thick-skinned dried grapes, by the intense heat of the sun. A *vin léger* is a fine limpid wine, not rich in alcohol, colour, and body, but yet not feeble and feminine; its constituent parts are in a harmonious equilibrium, without undue acidity or meagreness.

Nerf is the quality of wine which enables it to travel and resist changes of weather. It combines robustness, body, and vivacity. Wines are said to be *fumeux* or *montant* when they readily fly to the head; they are *liquoreux* when they preserve their sweetness a long time; they are *mous* when flabby, insipid, and void of strength.

I must remonstrate humbly with the œnologists of France, for what I conceive to be a misuse of the terms *vineux* and *vinosité*. They use them in the lower sense of *alcoholicity*; thus, *le vinage* is the addition of alcohol to wine. I respectfully contend that these words, and their English equivalents, *vinosity*, *winey*, and *winelike* should be used to measure that quality whereby wine is known from other liquids—the wine body and flavour *par excellence*.

To this collection of phrases many more might be added. One point which will strike the critic is the extreme difficulty there is in describing a simple sensation. In most cases it is necessary to resort to analogy from the sense of touch, so wines are called silky, velvety, marrowy. But if we cross-examine M. Maurial, and ask, what is *velouté?* he tells us it is *soyeux* and *moëlleux, avec corps et finesse*. Then what is *moëlle?* Why, it is that which gives the *velouté!*

CHAPTER V.

Importance of spirit to the revenue—Natural proportion of spirit in pure wines of all countries—Mulder's belief in natural port—Government inquiries and reports—Why spirit is added to wine—*Le vinage* in the south of France.

FERMENTED liquors of course contain alcohol. But alcohol is subjected to a very heavy tax, if distilled from the fermented liquor containing it. The revenue arising from the tax on distilled spirits in the United Kingdom for the year 1871, was 16,689,406*l.* When it was proposed to reduce the old heavy duties on wine, it was felt that if natural wine were let in at a low rate of duty, there was no reason to exempt *spirits* that might be mixed with wine: hence it became necessary to know what is the quantity of alcohol generated in natural wine.

It is surprising now to see the confusion which reigned on this subject thirty years ago; for instance, in the earlier wine analyses, as those by the eminent chemists Brande and Sir R. Christison, we find a long list of wines with their alcoholic contents arranged in tabular form; beginning with *Port*, the alcoholic contents of which are stated to range from 18 to nearly 24 per cent. (double those figures give an approximate result in *proof spirit*); *Madeira* comes next with the same strength; *Sherry* figures as the wine next in strength—viz., from 17 to 20; then come a miscellaneous lot of sweet wines, as *Constantia, Lachryma*

Christi, *Lisbon*, and *Malaga*, which range from 15 to 20; *Roussillon* and *Syracuse*, nearly the same; *Amontillado* figures at 16; then comes a sudden drop. The wines of *Bordeaux* and *Burgundy* figure at 9, 10, and 11. It was conceived that the high alcoholicity of Port wine was due to some law of nature, and this was the opinion of philosophers as well as of the vulgar. Thus the most eminent Dutch chemist, Mulder,[*] quotes from a writer named Gingal, who says "that genuine Port wines never contain more than $12\frac{3}{4}$ per cent. pure alcohol." Mulder does not believe Gingal. "How is it," asks he, "that all who have analysed Port wine have found from 17 to 21 per cent. alcohol? Is there no wine, except such as is adulterated with alcohol, exported from Portugal? And does Port wine, which is recognised as the strongest wine in the country that produces it, really belong to those not very strong wines which only contain 13 per cent. alcohol? For my part," adds Mulder, "I hesitate to accept Gingal's statement, although his experiments were made in Portugal." We now, however, know only too well that Gingal was right and Mulder wrong. But less than twenty years ago there were many who looked on Port wine, with its 40 per cent. of proof spirit, as a kind of natural product, which the Almighty had been pleased to create as the natural food for freeborn Englishmen. "*Nous avons changé tout cela.*"

There were two reports presented to Parliament—one called "Extracts of any Reports of an Inquiry under the Authority of the British Government in the year 1861 into the Strengths of Wine in the principal

---

[*] The Chemistry of Wine. Edited by H. Bence Jones. Lond. 1857, p. 187.

Wine-growing Countries of Europe;"* the other bearing the title "*International Exhibition.* Report to the Commissioners of H.M.'s Customs of the Results obtained in testing Samples of the various Wines exhibited, with a General Abstract of their Average Strengths, &c."† The tale they tell is unanimous and unmistakable. It is that the quantity of alcohol in *pure wine* may in round numbers be assumed to be 20 per cent. of proof spirit. Mr. Keene tested 569 samples at the International Exhibition, from Italy, Germany, Australia, and France; and in the following year the English Commissioners of Customs sent representatives into all the wine-growing countries of Europe to collect undisputed specimens of natural wine from the cellars of the original producers before any spirit whatever had been added. The fact remains, that of the 569 samples of liquid sent to the International Exhibition as *wine,* from France, Italy, Germany, Austria, and our own colony of Australia, the average—all, in fact, but a few exceptional specimens—yielded 18 to 22 per cent. of proof spirit.

These wines, be it observed, were sent by the growers. The evidence they give is corroborated to the utmost by those samples which were fetched and taken at the places of growth by the Assistant-Surveyors of the English Customs, as mentioned above. These samples were in every case authenticated as natural fermented juice of the grape, not mixed with any additional spirit; the figures which indicate their strengths in proof spirit are as follow:—17·75 per

---

* Ordered by the House of Commons to be printed, April 29th, 1862.
† London, 1863, No. 6448.

cent. for Bordeaux; 21·5 for Burgundies; 22 for Rhine wine and Hermitage; 24·3 for wine from the department of the Gard in the South of France. The average of Rhine wines, or Hocks, was 21·9, and of Hungarian the same. These figures are to be taken with the allowance that exceptional wines were met with, and particularly white wines, which ranged from 25 to 30.

France, Germany, and Hungary are the countries from which *wine*—natural wine—is procured. Now let us turn to Spain, Portugal, and Sicily. Let us see what is the percentage of spirit in the wines of those countries, according to Mr. Bernard:—

	Natural Wine.	Wine slightly fortified.	Wine fortified for English market.
SPAIN:—			
St. Lucar, Vino Fino, Sherry	27·0	—	—
Xeres, Sherry	27·2	30·7	—
St. Mary's, Amontillado	—	29·4	35·7
Montilla	31·7	—	—
Valdepenas	27·0	—	—
Valencia	27·2	—	28·6
Benicarlo	23·9	—	31·3
Alicante	28·9	—	—
PORTUGAL:—			
Port (average of 9 samples of natural wine from different growers)	23·5	—	—
Port slightly fortified	—	33·6	—
Ditto for English market	—	—	35·4

Respecting the wines of Italy and Sicily, no sample was procurable of Marsala, but all other evidence shows that the wines reputed the strongest contain only 20 to 22 per cent. of proof spirit; all beyond are mere exceptional specimens.

The English Customs fix 26 per cent. as the highest standard of alcoholicity in natural wine; and for ordinary purposes 20 per cent. may be taken as the average, yet there is no fast and hard line in nature. Mr. Griffin gives 29·20 as the percentage of some Rudesheimer, which I cannot believe to have been fortified. The powerful red wines of Australia, grown by Mr. Auld, Dr. Kelly, and Messrs. Wyndham, are sometimes above 26; let it never be forgotten, too, that white wines are as a rule stronger than red; but M. Terrel des Chênes,* speaking of 1865, the best wine year in France for many a day, says that none of the wines from the most sunny regions of France exceeded 14·10 degrees of alcoholic strength (equal to about 29 proof spirit).

The fact remains that many wines are fortified, as Port is, to a pitch nearly double the natural standard, and a good many other wines are fortified, though not to the same extent. And the question is, why?

The reasons are twofold: first, because the wine is not well made, and has not nerve enough to keep and travel; or, secondly, to please the vitiated taste of the consumer. Travelling implies movement that stirs up all sediment,—exposure to cold which precipitates soluble matter makes the wine thick, and so creates pabulum for fermentation;—alternated with heat which sets ferment a working, and leakage which allows access of oxygen.

The practice of *vinage*, or fortification, in France is chiefly practised in the south, where seven departments formerly had the privilege of paying no duty on the brandy that was used for this purpose. The practice is vividly described and denounced by M.

---

* Le Vinage. Pages détachées d'un questionnaire de l'Enquête Agricole. Paris, 4, Rue Neuve de l'Université, 1867.

Terrel des Chênes in a tract which is an admirable illustration of the moral side of wine culture. M. Terrel des Chênes shows that the departments of the East, West, and Centre of France claim no privileges, though their climate is colder and wetter; for during the period of activity of the vine, the South has daily some 3·8 degrees C. of heat daily, and 62 per cent. less rain; yet it complains that it is obliged to fortify its wines, because, first, they have too much sugar; and, secondly, have not enough alcohol. M. Terrel des Chênes describes much of the common vine culture of the south of France as abominable. They plant the vines which yield the greatest amount of the poorest wine; the vines are allowed to grow rambling, so that sun and air cannot reach the grapes; these cannot ripen; when the vintage comes, green grapes and rotten grapes are thrown in pell mell, " to ferment if it can, and sour if it choose;" and this is the liquid which they claim to fortify clear of duty; whereas, he says, a Bordeaux wine, even though no stronger in bad years, requires no fortification. If, he continues, the wines of the south are too sugary and not alcoholic enough, it is because they do not take the trouble to convert the sugar into alcohol; and any premium on *vinage* would be a premium on the ignorance, carelessness, and laziness which blight the processes of viticulture in the finest part of France.

Dr. Guyot, in his before-mentioned report, speaks of the wines of Roussillon as abundantly endowed by nature with colour and strength, and deprecates the addition of alcohol, the depraved taste which demands it, and the want of true commercial morality in the merchants who deal in it; and M. Maumené, as well as M. Terrel des Chênes, affirms that wine may be made without any addition of alcohol. If the juice is really too concentrated to ferment itself dry, water it, and gather the grapes before their juice is so concentrated; if too thin, add sugar, and let it ferment with the grapes, because then the fermentation will be more likely to produce an alcohol and ethers like the natural. But the spirit of wine which has been produced by distillation at a high temperature contains products not natural to wine, and the spirit of beetroot, of potatoes, and of maize contains many undesirable flavouring matters, and perhaps some day, says M. Guyot, "thanks to the poisons invented by chemistry, wine will be adulterated with spirit of turpentine or pitch, or with naphtha and spirits of bitumen, until the lake of Gomorrha will have nothing to envy."

Candour compels me to quote two writers of the highest repute who defend the practice of fortification. One is M. Thénard the eminent chemist, quoted by de Vergnette-Lamotte,* who asserts that it betters some wines that are thin and acid, (which if too feeble in sugar should, according to sounder policy, have been sugared before fermentation,) and that it enables the wines of the South of France to keep, (a thing better effected by more thorough fermentation, as Guyot and Maumené and Terrel des Chênes say). There is no doubt but that alcohol disacidifies wines by precipitating cream of tartar, and combining to form ethers with free acids; it checks fermentation, and suppresses parasitic vegetable growths. The other author who recommends fortification is the Rev. Dr. Bleasdale,† to whom I would point with admiration as

---

* Le Vin, &c., p. 107. † On Colonial Wines, Melbourne, 1867.

a patriotic and scientific œnologist, though I venture to doubt his soundness on this point. He advocates the throwing in of 1, 2, or 3 per cent. of very strong brandy, towards the close of fermentation, in order to fix and nullify the remaining albuminous matters and preserve some sweetness in the wine. But the Reverend Doctor takes Sherry and Madeira as the types to be aimed at, instead of a pure natural wine.

From a view of the whole matter, and from experience of the effects of various wines I would say, that *natural* wine is above all other to be preferred. Fortified wine contains fermentible matter in check; and this is believed from experience to be one potent cause of gout. The spirit used to fortify wine is liable to impurity, and if pure decreases the natural wine taste and substitutes mere heat. Yet it must be confessed that after a long *time* the added spirit seems to amalgamate and soften down, and as in the finer sorts of sherry, goes along with the production of those *old wine* flavours of which Madeira gives some of the best examples.

CHAPTER VI.

On the acidity of wine—Diseases of wine: M. Pasteur's treatment—Deposits in bottled wine—Beeswing, crystals, parasitic growths.

HOSE things are called acid which redden litmus paper, or which neutralize an alkali, or which give a certain impression to the tongue known as *sour*. Acids may be inorganic or organic. Amongst the former, the sulphuric, hydrochloric, nitric, and phosphoric are articles of diet or medicine; amongst the latter, the citric from lemons, tartaric from grapes, oxalic from sorrel, the acetic a product of sugar, the malic, racemic, &c., which exist in fruits, the tannic or astringent, and the lactic in sour milk. All acids without exception are poisonous in very large doses and concentrated forms; and some are extremely noxious in very small quantities, especially those which arise in rancid fat, in the fermentation of grain, &c. These, and not vinegar, are the acids of heartburn and gout.

Acids are greedily sought for by many persons, and avoided by others. The persons who seek them are usually the young, strong, active, and hearty, with free open pores of the skin, and good appetites. Acids do to the palate and stomach what soap and towels do to the skin—*i.e.*, they strip off its coating, make it redder, more active, and ready to secrete. Hence the love for lemon-juice, vinegar, and pickles at dinner, and the charm of acids to persons in certains kinds of bad health, torpid liver, coated tongue, &c. The secretions of sore throats are alkaline, and an acid liquor wipes this off, and leaves the surface clean. The persons who avoid acids are usually the torpid, and those with red tongues or skins locked up. The power of taking acids, however, does not depend on the dose of acid *per se*, but on its combination. Thus a very small dose of acid, strong and naked, might be intolerable; whilst almost any quantity may be taken if properly veiled, as it were, by sugar, extractive, gelatinous or fatty matter. Acids and these gelatinous matters are complementary to each other, and each renders the other wholesomer. Nevertheless, there are some persons who cannot tolerate any acid, whether naked or combined.

Should a man in good health be afraid of acids?—No more than he should be afraid of cold baths and brisk exercise. Some unlucky people can't take a cold bath without rheumatism, or a breath of cold air without bronchitis, or a long walk without exhaustion, or a cucumber without the colic. But are the healthy population, therefore, to avoid all that is cool and bracing? Certainly not; and so they should use that form of drink which suits an active, perspiring skin, and hearty supplies of meat. The stomach of a young girl should not be treated like an old woman's, which can digest nothing but bread and meat and alcohol. To keep the skin rosy, fresh, and young, the diet must not be that which suits the faded, withered, torpid skin of age, " colore mustellino.". The history of scurvy in the navy should also be borne in mind, and the number of skin eruptions and blood disorders for which the combinations of potass and vegetable acid found in wine are prescribed by the physician.

Nevertheless, wine that is too sour to be pleasant should be shunned, because it offends that divinely-ordained instinct of taste which teaches us what is good for us.

The acid in wine is first, that of the grape which remains after fermentation; second, that generated during and after fermentation, partly by oxydation from the air, *pure and simple*, partly by the presence of parasitic growths, or by the presence of decaying organic matter.

In good wine the acidity is due to tartaric, malic, and volatile acids, each wholesome *per se*. If too acid, the fault may be excess *simpliciter*, or more probably defect of *body*, which should veil the acid. The only test of *quantity* of acid is the chemical one; and this shows that very first-class wines of the Rhine and Moselle, contain most acid; port and sherry least; but it must be remembered that one-fifth or more of port and sherry is not wine, but spirit; and secondly, that the makers of sham wine can put in as little as they like, or can neutralize natural acidity by chalk or plaster: hence quantity of acid is no test of quality of wine. Nay, the tartaric and other organic acids may actually themselves decompose, and spoiled wine be less acid than the same wine sound, as in some wine diseases.

Hitherto we have spoken of pure dry wines. We need not trouble ourselves about the acidity of sweet wines—*i.e.*, those containing much sugar; nor about fortified wines, which cannot be very acid because of the quantity of spirit added, and whose characteristics are enormous alcoholic strength with the sweetness of unfermented juice.

The second cause of acidity is exposure to the air, and the absorption of oxygen. A third cause is the contact of decaying organic matter. This is well exemplified by No. 22, in the following table.

Any putrid taint may sour meat, milk, wine, or man. Dead flies, says the book of Ecclesiastes, cause fragrant ointment to stink. A few dead flies, now, in the East, will cause any jar of honey to ferment and acetify. And the same will happen to wine. Out of three samples, Nos. 21, 22, and 23, drawn from the same cask, one was put into a bottle into which a few flies got when the cork was left out. This small quantity of decomposing organic matter converted it into vinegar, besides generating other acids (concerning which, consult Mr. Griffin).

Excess of acid is a common fault with wine of the extreme northern limits, and in cold seasons. The grapes do not ripen; they contain absolutely too much acid, and too little sugar. Therefore they cannot of themselves be rich in alcohol, and alcohol it is which most efficiently, within certain limits, covers the taste, and neutralizes the effects of acidity. The observations on this point made by Mr. Griffin (*op. cit.* p. 109) deserve the greatest attention. In good wine "the weight of the alcohol should have a certain relation to that of the acid." Of course there must be a certain *quantity* of each; but this being granted, the proportion must be within certain limits.

"Light wines," says Mr. Griffin, "have 450 grains of acid in a gallon, of which almost one-fourth part is volatile acid, and three-fourths are fixed acid. 2. The quantity of absolute alcohol is twenty-one times as much as that of the total acid by weight. 3. There is no sugar. These conclusions are founded upon the opinion that the acid is the prime regulator of the tastes of wines." With an acidity of 450 grains of acid in a gallon, you can have a first-rate wine, but the acid must be covered with from 20 to 25 times— with at least 20 times—its weight of alcohol. As to sugar, it is a superfluity. The average quantity in 21 light wines was under 150 grains per gallon, while the acid had a mean weight of 417 grains, and the alcohol a mean weight of above 7000 grains. The sugar residue seems to be the mark of failure in perfect fermentation. There ought, says Mr. Griffin, to be none left in the completed wine, and in many first-rate wines there is none.

The following table of the acidity of certain wines was published in my former edition, as given me by Mr. Griffin, whose general conclusions as to quantity of acid are summed up thus:—

1. Good wine contains a quantity of acid that is equivalent to from 300 to 450 grains of crystallized tartaric acid in a gallon.

2. Wines with less than 300 grains of acid in a gallon are too flat to be drinkable with pleasure.

3. Wines with more than 500 grains in a gallon are too acid to be pleasantly drinkable.

4. Wines with more than 700 grains in a gallon are undrinkably sour.

The figures mean this—Voeslauer wine is as acid as a liquid containing 375 grains of tartaric acid per gallon. And so of the rest.

Of these wines, Nos. 1, 3, 5, 7, 8, 13, 16, 20, 21, and 23 are good; Nos. 5, 7, and 16 very good; and 24 first-rate. So that at any rate what suits most palates has acidity equal to 73 grains of tartaric acid per bottle.

Nos. 2 and 14 were old, spoiled bottles; the palate detected an undrinkable *quality* of acid; yet they are not intrinsically very sour. No. 12, a *sour* claret, see p. 54, yet not so acid, chemically, as No. 16. Nos. 3, 4, 5, 6, and 23, show examples of wine becoming less acid by age. Nos. 10 and 18 are fortified.

In the *total* acidity is included the *fixed*, which Mr. Griffin believes to be pretty constant in quantity, and the *volatile*. The volatile varies much. It is high in good wines, and consists in the fragrant complex ethers generated by the tartaric acid; and it is *very* high in wines that have much acetic acid, and other volatile noxious acids. The volatile acidity in Nos. 1, 7, and 8, good wines, was equal to 45, 75, and 85 of tartaric acid per gallon respectively. That in 3 and 4, to 128, and 110; that in No. 22, to 938 grains per gallon !*

No.	WINES.	Total acidity.
1.	Voeslauer, Schlumberger's	375
2.	Szamorodny, spoiled	430
3.	White Capri, Fearon's	450
4.	Ditto 3 months opened	440
5.	White Keffesia, Denman's	350
6.	Ditto 3 months opened	300
7.	Ofner, Max Greger's	375
8.	Thera, very old, fine, and soft, Denman's	350
9.	Santorin, opened April 18, 1865, ditto	410
10.	Tarragona, sample from docks	325
11.	Fronsac, open 5 months, too strong and sour at starting	500

* N.B. These estimates of acidity agree with those of Liebig, Fresenius, Diez, and Günning, as quoted in Bence Jones's Trans. of Mulder.

No.	WINES.	Total acidity.
12.	Fronsac, purchased April 18, 1865	400
13.	Claret, 21s., poor and thin	340
14.	White Diasi, soured	490
15.	Como, 1862, sample, Denman's	400
16.	Good ordinary Claret	450
17.	Dioszeger Bakator	450
18.	Oxford Sherry, at 36s.	325
19.	St. Elie	475
20.	Gilbey's Castle I Hock, at 16s.	440
21.	White Keffesia, *ex* 'Ada,' March 28, 1864; sample drawn March 1864	380
22.	Same wine drawn at same time, in bottle with *dead flies*	1300
23.	Same wine, drawn November 4, 1864	375
24.	Rudesheimer, 84s.	440

Wine is a liquid exceedingly greedy of oxygen. But the most effective instrument of oxydation is the growth of the *vinegar plant*, mother of vinegar or *mycoderma aceti*. And this leads me to say a few words on the parasitic vegetables found in wine, and the diseases they occasion.

M. Pasteur's theory of the nature of change of wine in time is, that it is brought about by the slow limited action of oxygen on wine in cask and bottle, which *oldens* the wine, and causes its colouring matter and tannin to become brown and insoluble, and deposit a crust. Hence the wine in time perishes and dies of old age.

New wine in vessels partially filled becomes covered with a thick mould of *fleurs de vin, mycoderma vini,* a microscopic parasitic plant, composed of little corpuscles arranged in bead-like strings, which Pasteur believes to feed on the wine, and transform its alcohol into water and carbonic acid. Far different is the *mycoderma aceti,* a similar plant, but much smaller, which converts the alcohol, by imperfect oxydation, into acetic acid.

When wine is infested with the germs of this plant, it is never safe in cask or bottle. Some districts are infested with it, whilst others are not; and it is to avoid this that the great care is needed, which is used to keep wines from the contact of the air and disturbance of their lees, when they are being racked off, or bottled.

After acescence, M. Pasteur describes a wine malady in which the wine is said to be *tourné, monté, avoir la pousse,* &c. The wine becomes flat, sour, and turbid, and if moved in a glass held to the light seems as if it contained a something silky, waving to and fro. It gives off bubbles of gas, which sometimes threaten to burst the cask. This, Pasteur believes to be a special fermentation allied to the *lactic,* but producing abundance of vinegar, and due to the development in it of a number of microscopic filaments of the extremest tenuity, often less than 1-1000th part of a millimetre in diameter. It exists in the wine from the first, but assumes a rapid development during the first hot weather of the next year.

A third malady is when wines, especially white, lose their limpidity, become flat and insipid, and pour like oil. This, the *maladie de la graisse,* is owing to another parasite growing in filaments, differing in some microscopic respects from the last. The addition of tannin is a good remedy.

Yet a fourth disease—with which I am very familiar, and which has ravaged some very fine wines, especially Volnay and Assmaunhauser. This is the malady of bitterness;—*amertume, maladie de l'amer, du goût de vieux.* It is easily recognised thus: invite a man to a friendly dinner; begin well with a little glass of old Madeira; next, a glass of some appetizing light wine; then, whilst he takes his slice of mutton, invite him to take a glass of Volnay or Assmannhauser, and let him be helped first. It is the critical point in the dinner; hunger beginning to be appeased, and the palate in its highest state of receptivity. You watch the guest as he sips the wine during a pause in some joyous talk. That which was satisfaction in his demeanour, should be ecstasy. But no! the thermometer of his visage sinks ten degrees; he hesitates; looks sad. Can it be that he is ill? Alas, no; you turn to the glass poured out for yourself, and there, instead of the bright, ruby, delicious mouthful, you have, as it were, a prematurely old, thin, yellowish-brown, sad liquid, with no taste of wine, no alcoholic strength, only a flat bitterness. What are you to do? First, resolve that you will never offer a guest an old red wine out of the bottle, without preliminary examination and decantation. Secondly; give him the best substitute at hand. Thirdly, descant learnedly on the disease which you may truly affirm to be particularly fatal to the grandest wines, one of which you have exhumed in his honour. Fourthly, you may improve the occasion by corking up the wine, and examining its sediment with a fine microscope after dinner. Lastly, you may console yourself by the reflection, that now you have learned *why some Sherries are bitter*. But this we must speak about presently.

With the microscope we find three things in wine sediments: First, the crystals, of cream of tartar and tartrate of lime, which may be seen by the naked eye upon the cork, if wine has been bottled rather new. Secondly, the colouring and extractive matters which ought to be deposited as beeswing, lining the bottles, in thin, filmy, tenacious crusts, looking under the microscope like an organized granulated membrane. Such crusts I have seen, par excellence, in 1820 and 1834 port, in Santorin and in Naussa. They generally adhere to the bottle; but if shaken up, they subside directly, leaving the wine clear and bright as before. Other deposits of this material are powdery, light, easily disturbed, and slow to settle again, and make the wine sour and bitter till they are got rid of. The third portion of the wine sediment consists of the remains of parasitic growths. These, says M. Pasteur, from their tendency to mix with the wine, are the occasion of great loss when the wine is racked off or decanted. It is rare, says M. Pasteur, to find any red table wine, grand or common, entirely free from parasites in its deposit—especially the filamentous mass of the *maladie du vin tourné*.

The existence of these parasites, as we have said before, explains many of the rules which "old experience" has laid down for the preservation of wines, especially fining, freezing, and filtration. Besides, there is Pasteur's practice of *heating*, or *chauffage*, which consists in exposing wine, in cask or bottle, for a longer or shorter time, to a temperature of about 45°–50° Centigrade, or 110°–130° F. This may be done in a chamber by stoves or by steam, and whether the wine be in cask or bottle, it seems to kill the parasites, and give the wine softness and maturity, without alteration of colour or of taste.

The ensuing woodcuts will give the reader a rough idea of the nature of the sediments in wine, viewed microscopically. Each shows the sediment of wine opened August, 1872.

As practical deductions I venture to say—1. That the deposits which are least incompatible with a healthy state of wine, are the heaviest, go to the bottom most quickly when disturbed, and leave the wine clear. Such are the crystals found in the best wines. 2. Next to this in innocency is a compact crust. 3. That a red wine with a compact crust may decay, thin, and go utterly sour from age, without parasitic intervention. 4. That spoiled wine often abounds in parasitic growths. 5. That some wine may abound in parasitic growths, of the oval-bead shape, and be quite turbid on the least disturbance, and yet the wine filtered off may be bright, sound, and exquisite. 6. Lastly, but for all this, when a man sees grumous powdery deposit in wine, red or white, and especially of a filamentous character, he had better use the wine than let it decay.

The woodcuts show the deposits of—

1. *Haut Sauterne*, 1851, p. 74. Little save minute bundles of crystals of cream of tartar.

2. *An ordinary St. Julien*, 1858. Thoroughly sour from age. A good firm beeswing, studded with abundance of little molecules, which Pasteur declares not to be sporules, but only petty molecules of oxydized extractive.

3. *Old White Dry Ruszte*, of Max Greger's, in my cellar since '65. When filtered, a splendid Madeira-like wine. A very large deposit of the minutest light sporules? Shows also a fragment of cork.

4. *Volnay, E. Laussot*, 1858. A wine decidedly aged, attenuated, tawny, bitterish, but uncommonly good old wine on its last legs. Some firm crust, and an infinity of minutest threads proceeding from oval sporules. The wine decants quite clear.

5. *Wine from the Loire*, quite soured, but bright;

some crust; crystals of tartrate of lime, and abundance of sporules; *maladie du vin tourné?*

6. *Vaucluse;* intensely sour and muddy; large quantity of filamentous stuff. *Maladie du vin tourné.*

7. *Red Bukkulla,* from Wyndham, of Dalwood. A stout, admirable wine. Peculiarly shaped crystals.

8. *Falerno.* Sour, and quite decayed. A large quantity of filamentous stuff, like that described by Pasteur as the cause of the *amertume,* and of the *vin tourné.*

CHAPTER VII.

Classification of wines—Light or pure wine—Sweet wine—Sparkling wine—*Vins de liqueur* or fortified wines—Geographical classification—Wines of France—Political finance and prohibition—Bordeaux and the poet Ausonius—Classification of Bordeaux wines—Médoc—Graves—Petits Graves—Sauternes—Libourne—St. Emilion—Bourg—Blaye—Entre deux mers—White wines—Red wines—Nomenclature of wine—Medical uses—Different kinds of thirst.

OW we must hasten on from wine in the abstract, to the various kinds we meet with in actual life. But first of all, we must say three words on classification, whether as to the nature of the wine, or the place it comes from.

Four kinds of wine are met with—first the real or pure wine, grape juice fermented dry, of which we may take red Bordeaux wine as our pattern. Secondly, sweet wines, containing a considerable quantity of unfermented sugar, but not fortified, as the Tokay. Thirdly, sparkling wines, like champagne; and fourthly, fortified wines, of which port and sherry are common examples.

Wine is generally named after the place it comes from; and speaking broadly, the wine of every country has some qualities of its own by which an experienced person can recognise it. Of the several classes in ordinary use three are French,—the Bordeaux, Burgundy, and South of France,—besides the sparkling wines of Champagne.

To the east, the Moselle and the Rhine, with their

tributaries; Switzerland, Austria, and Hungary; to the south, Greece, Italy, Spain and Portugal, Sicily, and Madeira, the Cape of Good Hope and the Australian Colonies, each supplies wine which we must say a few words upon. Draw a line from the mouth of the Loire to the point where the Rhine enters the swamps of Holland, and you have the northern limits of profitable vine culture in Western Europe. In few words, it is neither the heat of the tropics nor a rainy climate like that of England that the vine requires. Like a hardy plant, it bears a considerable amount of cold, but it needs a long hot dry summer to ripen its fruit. Wine is made in Palestine, Egypt, and Persia, but hardly as an article of commerce. From Peru I have tasted a sample of Pisco; but am ashamed to say that I know the wines of North America only by repute.

Without any doubt wine can be made in England in some favoured spots. In Dean Hook's Lives of the Archbishops of Canterbury, some records are cited of ancient vineyards, such as those of the Bishop of Ely, on the western slopes bounding the valley of the Flete. Wherever those much maligned men "the monks" settled, there they must have taken wine for the sacrament, and the effort to plant the vine as an instrument of civilization. Wine is still made in the neighbourhood of Guildford, for private use. But it is evident that our forefathers soon gave up the idea of home supply, and that their wine was procured from those countries in the west of Europe which were most accessible by sea. Chief among such sources were Gascony, Guienne, and Acquitaine; whilst rarer wines came from the coasts of Spain, from the Rhine, from the Canary Isles, and from Cyprus.

approachable Châteaux Yquem and Lafitte, to the good sound ordinary which we can get at a shilling a bottle. The general character of the Bordeaux wine is purity, subastringency, lightness, and fragrance.

I may say a few words on the white wines first. *Ceteris paribus*, white wines are more perfect of their kind than red; their flavours are finer; they are more seductive, subtle, and feminine, more stimulating and expergefacient (or the contrary to narcotic). They are for those who take for their motto Ὀυ χρὴ παννύχιον εὕδειν. They veil and neutralize that which is fat and glutinous. They act as foils to the taste of soup, fish, and the rich dishes which a hungry man likes to begin dinner with, better than red wine, and they are more aperient, or at least less astringent. They are not so popular nor commonly used, except by sensible people, who are always a minority.

The white Bordeaux wines vary from the thinnest acidulous disembodied wine to the lordly Sauternes. Of the Château Yquem I have spoken elsewhere as a wine of perfection, in which a large quantity of the richness of the grapes becomes blended by age into a body of inexhaustible fragrance. I open a bottle from MM. Fauché, Fils, and C. Brisac, of "Haut Sauterne, 1851," consequently twenty-one years old. It is of the palest primrose colour, distinctly acidulous, subtle and penetrating, soft, and without a particle of heat, little or no body, and with a fine floral bouquet. The price was only 5fr. f.o.b. at Bordeaux in 1865, and it shows how little alcohol is needed for preservation if there is purity. I have before mentioned a *Langoiran* from the same house. I have notes of a Barsac, 1859, from the same, at fr. 2.50; of a Sauterne, sound and pleasant, from Trapp, at 20s. per dozen; of

a wine from the commune (or parish) so called; a "Château Margaux," means from the vignoble of a special estate. Pauillac is the name of a canton in which Château Lafite is situated. Other familiar names are St. Julien de Reignac and St. Estèphe.

The second Bordeaux wine district is Les Graves, also on the left bank of the Garonne, above the Médoc and continuous with it. The chief wine here is the Château Haut Brion, one of the first class.

The third district, *les Petits Graves*, higher up the left bank, produces good red and white wine. Here we meet with the well-known name Podensac.

The fourth district, still extending up the left bank of the river, is the great white wine district, whose produce has the general name of *Sauternes*, of which the queen is the *Château Yquem*, also more frequently quoted than tasted. Barsac, Bommes, and Château Suduiraut are other seductive names.

Next to the Château Yquem stands the Château Vigneux Pontac, of which the 1861 wine was adjudged superior to any Rhine wine at the French Exhibition of 1867.

For the fifth district we have Libourne, including the St. Emilion and Fronsac districts, on the right bank of the Dordogne, and that of Cubzac.

Sixthly, there are the districts of Bourg and Blaye on the right bank of the Dordogne, while *Entre deux mers* occupies the angle.

From the district of which I have given a sketch, in which about 2304 *principal* wine growers, and about 1584 *principal crûs* of wine are enumerated in M. Féret's work, and in which between two and three million hectolitres of the best wine in the world is made in most years, there are all qualities, from the un-to whose work* I must refer any one who desires more information, we find the Bordelais territory subdivided into the following parts:—1st, The Médoc, upper and lower. Draw a line from Arcachon to Blanquefort, and you have a triangular bit, bounded on the south by this imaginary line, on the west by the sea, and on the north and east by the river Gironde ; and it is on a strip about eight kilometres wide, along the left bank of the river, that the celebrated wines of the Médoc are produced. Here we read reverentially of the five grand classes, or *crûs* (more often talked about than tasted by ordinary mortals), the first class embracing Château Lafitte, Château Margaux, Château Latour (and Château Haut Brion in the Graves); the second class embracing sixteen wines, such as Mouton, Rauzan, Léoville, Pichon Longueville, Cos d'Estournel, &c.; the third class, thirteen wines, including Kirwan, Lagrange, Langoa, &c.; the fourth class, St. Pierre, Château Beycheville, &c.; and the fifth class, Pontet Canet, Batailley, &c., &c. These wines, according to their year, their age, &c., fetch from 1200 to 10,000 francs for barriques of 225 litres. Below these five grand *crûs*, the Médoc wines are ranked as *bourgeois* of two or three degrees, *artisans* and *paysans* ; below these come the *ordinary*.

A distinction is made between wine from the *Côtes* or hill-sides, and wine from the *Palus* or marsh.

There are forty-two communes in the Médoc in which wine is made, and from each of which wine takes its name; though be it observed that the *grand* wines are named after the *estates*. A "Margaux" wine means

---

* Bordeaux et ses vins, classés par ordre de mérite. 2ième Ed. Paris : V. Masson, 1868. See also, Maurial, *op. cit.*

no longer obtain these wines, drinking-songs ceased. My friend arrived at the conclusion that when the English used to drink French wine, it made them sing and be merry; but when they began to drink Port and Sherry, these made them stupid and brutal."

The wines of Bordeaux have a reputation of no late date. The poet Ausonius, who was a native of the city, and late in life its prefect, is believed to have lived at what is now the Château de Bel Air, in the district of St. Emilion, in the arrondissement of Libourne. He lived between Anno Domini 320 and 400. In his idyllic poem in praise of his country place, or *villula*, which his great-grandfather had possessed before him, he says—

> "Agri bis centum colo jugera: vinea centum
> Jugeribus colitur, prataque dimidium—
> Sylva supra duplum quam prata, et vinea, et arvum,
> Cultor agri nobis nec superest, nec abest—"

In speaking of the Moselle, in another idyll, he says it reminds him of the bright Bordeaux—

> "nitentis Burdigalæ"

Again, in praising the oysters of Bordeaux, he says they are as good as the wine—

> "Non laudata minus, nostri quam gloria vini."

What we in England call the Bordeaux District is, according to the modern subdivision of France, the Department of the Gironde. If we look on the map we see that within its borders is situated the confluence of the Garonne with the Dordogne. The soil is very varied, hilly and marshy, but all tertiary gravel, sands, clays, marls, and alluvium.

Following the arrangement of M. Edouard Féret,

through their dryness, for any other crop," says M. Michel Chevalier, "become lucrative through the vine, and lands already fertile have their fertility doubled."*

The teetotallers who rave about the wickedness of converting a twopenny loaf into a pot of beer, are dumbfounded by the vine, which grows on mere scrub where no barley would flourish.

It really seems a stupid thing that two nations like the French and English should vex each other by putting fiscal difficulties in the way of supplying their mutual wants. But a truce to this reflection. Let us now reverently approach the wines of that great district which has the ancient Burdigala or Bordeaux for its centre; that which gave the ancient wine of our forefathers; that which inspired England when it was yet merry England. M. Terrel des Chênes quotes to the following effect from a speech of Cobden's, which I hope that great man's disciples will not ignore in any further alterations of the wine duties:—" Every nation except the English, considers the French wines the best in the world. We alone take adulterated wines in preference to them. Those of us who can get them, prefer those brutalizing and inflammatory mixtures called Port and Sherry. A friend of mine had lately a fancy for seeking material amongst our national ballads for a collection of drinking-songs. He told me that he found that all these songs were in honour of French wines—Champagne, Burgundy, or Bordeaux. They were all old songs, written in the times when our forefathers drank or preferred the French wines; but from the time when they could

---

* Quoted, Parfait Vigneron, 1867, p. 52.

have an honest English chop or steak." Now suppose a Frenchman were to enter a London eating-house and say as loudly as possible, " Waitère! Mistère Godam! don't bring me your *sacre* lumps of raw flesh and *légumes à l'eau*, but let me have an elegant French repast," he would run the risk of being kicked out. Now what is to cure the Englishman of this offensive bumptiousness, and give him a little of that consideration for other people, that true suavity and grace of manner for which the French are so distinguished? Why, give him some good Bordeaux wine! teach him that force is not the only virtue; and promote the intimate alliance of the two nations, so long separated by unnatural prejudices.

We need not dwell on the enormous production of wine in France. It is grown in seventy-nine out of eighty-nine departments, and the quantity produced was estimated at 50,456,421 hectolitres in 1864,* and at 70,910,220 hectolitres in 1869.† Besides this wine, which is valued at 1,628,807,753 francs, the *marc*, or pressed husks, yields 1,193,000 hectolitres of brandy, value 59,650,000 francs; the residue from the *marc* is valued at 80,000,000 francs for fodder of cattle, and 16,740,000 francs worth of manure. The young shoots of the vine are worth 23,860,000 francs as fodder, and the old wood at 95,400,000 for fuel; so that the whole value of vine products amounts to 1,904,457,753 francs, which is estimated to be equal to the support of a rural population of 7,617,828 individuals. No other crop anywhere is capable of maintaining so large a population. "Lands unfit,

---

* Quoted from Pasteur, Études, &c. p. 2.
† Quoted from Le Parfait Vigneron, for 1871-72, p. 143.

lation—at the end of the last century revolted against the descendants of the Chludwigs and their feudal lords (who were really of Franco-German descent), it is true that the depressed town-populations committed great excesses and most deplorable cruelties. But whether slaves or free, it seemed a religious duty with the country parsons and squires to hate the French, first for their slavery, next for their revolution. Everything French was painted in the blackest colours. Of course, free-born Britons could not tolerate French wine, thin, sour, and impregnated with mischief. Good Claret was described as willing to be Port if it could. People who then knew, and now know no more of France than a Frenchman would know of England if he should live at a London hotel and spend his evenings at Cremorne, take upon themselves to revile everything French; not knowing that the French landed proprietors are among the most virtuous, industrious, frugal, and intelligent persons on the face of the earth; their wives and daughters patterns of religion and morality; and the peasants frugal and self-reliant. The mass of the French are of the same blood with our fellow subjects in Ireland, the Highlands, Wales, and Devonshire. The two nations ought to mingle, and supply each other's defects, moral and material. If (to borrow some wine terms) the English have too much of the *corsé, dur, vineux, ferme,* not to say *âpre* and *montant,* they have need of a *coupage* with some of the *moëlleux, soyeux,* and *velouté* of the French. I was once at a restaurant near Paris, when an Englishman came in. The waiter approached him, and politely handed him the bill of fare. "Garsong," said our countryman, "don't give me any of your d——d kickshaws, but let us

France. Claret harmonizes with the light ethereal character of the great Celtic race. Port was a symbol of Whiggery and Presbyterianism.

> "Erect and firm the Caledonian stood,
> Old was his mutton, and his claret good:
> ' Let him drink port,' the English statesman cried.
> He drank the poison, and his spirit died."

These are lines from the tragedy of "Douglas," by John Home, who died 1805, whose grand-nephew, G. Y. Home, of Red Cross Street, Bristol, is one of those honourable and chivalrous wine merchants who regard the character of their wine much more than their profit.

The reign of Port coincides with the growth of the national debt, the isolation of the English from continental society, the decay of architecture, and that "chauvinism" (as it has since been called), that ignorant national pride, which begat the popular dicta that "all foreigners are fools," that "one Englishman can lick three Frenchmen," that true religion, and virtue, and freedom, and morality, and beef and pudding and good beer, were special blessings of the English, conferred on them by a discerning Deity; whilst Popery and tyranny and wooden shoes, arbitrary power and a diet of frogs and sour wine, were the especial curses of the French. Not to drink damnation to the French in Port wine would have been considered a treason to the House of Hanover. Look at Hogarth's prints, and read Smollett's novels, to see how the French were despised for their abject submission to the remains of the feudal system, which were, it is true, vexatious enough. Even the partridges were said to be lean because of the tyranny of the nobles. But when the people—the Celtic popu-

one per cent. of the wine used in England. Well may we say with Mr. Shaw, that the French wines have not had fair play. For political purposes, our people were bribed to drink of the "drugged chalice" of Portugal, whilst hatred against the French has been instilled as a religious creed, and their fine wholesome wines kept out by extravagant differential duties.

During the early part of the eighteenth century, French wine was banished from England by politicians; but the educated and intellectual classes grumbled as much then as they would now if all wines were banished save South African. As cheap wine now is called "Gladstone," so, but with less of respect, port was called "Methuen." The stage reviled it; the poets Prior, Shenstone, Pope, all had a fling at it as dull, muddled, humble, thick, flat, cheap stuff. I can refer to one besides, who was physician as well as poet—Armstrong, author of the "Art of Health," who, in describing a man's sensations on awaking, after drinking port over-night, says—

> "You curse the sluggish port, you curse the wretch,
> The felon, with unnatural mixture, first
> Who dared to violate the virgin wine."

Again, when speaking of wholesome wine, he praises

> "The gay, serene, good-natured Burgundy,
> Or the fresh fragrant vintage of the Rhine."

Again, he describes Burgundy as the drink for gentlemen, and port as an abomination—

> "The man to well-bred Burgundy brought up,
> Will start the smack of *Methuen* in the cup."

The political significance of the wine duties was unmistakeable. Claret was a symbol of loyalty to the ancient Royal House, which had taken refuge in

In Froissart's Chronicles* we read that when young King Edward the Third was at York with his army, in 1327, " good wines from Gascony, Alsace, and the Rhine, were in abundance and reasonable," whilst even farther north, on the banks of the Tyne, the country people bring a poor thin wine in large barrels, and sell a gallon for six groats, though barely worth sixpence.

France continued the great wine source for England till near the end of the seventeenth century. Then French wines were heavily taxed to spite the French, and, after some fluctuations, were visited with a duty almost prohibitory. The alliance of Charles II. with a Portuguese Infanta, the support given by the Court of Versailles to the Stuart family, and the intrigues of Louis XIV. in Spain, successively induced the English Government to cultivate a closer alliance with Portugal, which culminated in the Methuen treaty in 1703. In 1675, says Mr. Denman, there came to England 14,990 pipes of French wine to 40 of those of Portugal; in 1676, there were 19,290 French to 160 Portuguese; but between 1679 and 1685 only 8 pipes of French wine were imported, whilst 13,760 pipes of Portuguese came in.† By the end of the 18th century, French wine did not form

---

* By Johnes: Lond., 1839, vol. i. p. 17.
† Up to 1693, the duties on French and Portuguese wine were equal; having been raised since 1671 from 4d. to 1s. 4d. Between 1693 and 1697, the duty on French was 2s. 1d., on Portuguese and Spanish 1s. 8d. Between 1697 and 1707, French paid 4s. 10d., Portuguese 2s. This process went on from bad to worse, during the century and a quarter following, so that in 1813, French wine paid 19s. 8d. per gallon, and Portuguese 9s. 1d. In 1831 the duty was equalized, and made 5s. 6d. per gallon.

another, 1858, from the same, a delicious, full-flavoured, grapy wine. We may put these wines into two categories : the very full-bodied, subluscious, and grapy ; and the dry. The former are expensive, full of flavour, and deserve to be sipped leisurely ; the latter are decidedly not the wines for people who want to go to sleep. They excite the appetite, and rouse the heart and brain. I believe that these, like other white wines, are much too little in use.

Next for the audacious attempt to epitomize the qualities of the red wines of the Gironde in the short space available. For my purpose I may divide them into expensive wines, and ordinary or cheap wines, never forgetting that the original motive of my writing on wine at all, was to show that good wholesome wine can be had for moderate prices, and that the extended use of it would be a public benefit.

Respecting the more expensive and classed wines, I have mentioned *Château Lafitte* at a former page. Within these few days I have tasted a genuine *Château Margaux* of 1848, a wine that has passed its climax, getting thinner, and very slightly bitter ; but what cleanness and flavour ! The *Cos d'Estournel* I used to drink at the hospitable table of S. P. C., whose memory is as fragrant as his wine. The *Château Mouton*, 1861, at 7fr., from the Caves de la Gironde, a finished, delicate, charming wine ; Château Langoa, 1861, at 54s., a dry, light, most agreeable wine ; Latour, Battailey, Beycheville, Léoville, of all which I have tasted specimens of undoubtable genuineness ; these all bring back the memory of first-class wines. I believe no better rules to drink by can be given than my own ; so let me say that these wines have absolute oneness of taste ; they are generous without heat ; acid with no obtrusiveness of acidity; have not a particle of sweetness, a certain fine, stable, round taste, astringency perceptible but not obtrusive—but of the body, and of the bouquet, who shall presume to speak?—the exquisite full body, and the delicate perfume which is felt in deglutition, combined in a whole of exquisite softness and harmony, and adorned with that magnificent dark ruby carbuncular colour of unfathomable brilliancy. Such are the characters of the grand wines, which amongst the humbler classes, to whom I belong, must be reserved for occasional festivity and for the recuperation of the sick.

I have been much attacked for giving my lucubrations the title of "Reports on Cheap Wine." Doubtless the word "cheap" has, with some people, unpleasant associations. And it is quite true that there are some great luxuries which are of necessity scarce, and therefore fetch a high price, and that some cheap imitations of these are nasty enough. Yet our greatest luxuries consist really of things which are of themselves cheap enough, but which, if perfect of their kind, give us half the comforts of our life. The aim of the really luxurious man should be not to indulge occasionally in costly and exceptional delicacies, but to have all ordinary surroundings, all homely details, as good in their respective kinds as possible. What can be greater luxuries than pure fresh air, moderate warmth, and cleanliness? What a treat is good yellow primrose soap and a good towel? Yet how cheap compared with some perfumed soap that can't cleanse, and diapered towel that wont wipe dry! How absurd to see tasteless and second-rate bread, bad butter, and bad potatoes, even in houses where the expenditure in unnecessary luxuries is extravagant! where, perhaps, some port, Madeira, or "Château something," at fabulous prices, are the boast of the cellar, but where they can't pour you out a tumblerful of decent ordinary claret at luncheon. I affirm that the greatest luxuries are derivable from the enjoyment of cheap things—*i. e.*, of things produced bountifully and plentifully, when such things are good, each according to its proper standard.

This is pre-eminently the case with wine. There is far more enjoyment got out of wine by the many who are able to drink without stint a good pure ordinary, than by the few who are able to purchase the rarest vintages; and as, like all things for popular use, wine must be cheap if popular, it is worth while to examine into this point. Are we to believe the few who, in solitary grandeur, tell us that all cheap wine is rubbish? or can we persuade ourselves that wine is really plentiful, and that there is wine enough in the world for us all?

Of cheap claret, every variety is to be got, from that which would do for vinegar to that which recalls many of the properties of the grand wines spoken of above. If a man imports from a grower, and goes to work economically, he can lay it down at less than a shilling a bottle. I must reserve to another chapter what I want to say about bottling wine at home, but may observe that it is positively unfair to drink any wine that has been less than six months at rest in bottle. If it were the finest *Château Quelconque*, only just bottled, it would be heated, *éventé*, flat, sourish, destitute of flavour; how wrong, then, to expect it in perfection when it is had in by half-dozens fresh bottled? It may be very good, but would give three times as much pleasure if at rest six months.

There are several wine merchants who make efforts to obtain good wine direct from the grower, and to supply it at reasonable prices, and to some of whom I am indebted for specimens and for information. Amongst these I must quote Messrs. Trapp, whom I must thank for specimens of Lachenaye, at 36*s*., 1861; Château Langoa, 1861, at 54*s*. a charming light, dry wine; and Château Latour, 1862, at 70*s*. Messrs. Collier, of Plymouth, have sent me specimens of some of the finest growths of 1864. Mr. Blaxall, of Lamb's Conduit Street, some Château Latour Gueyraud, from the Palus of Entre deux Mers. Mr. Manley F. Bendall, Rue de la Verrerie, Bordeaux, sent me a most sensible letter, with samples of Médoc, Ladon, and Margaux, as types of the lower classes of pure and wholesome wines within the means of all customers. M. le Baron du Périer de Larsan is sending over wines remarkable for their purity, high quality, and moderate price.

But any wine merchants in London or the country, or at Bordeaux, can easily supply a *barrique* of good ordinary, old or young, light or full, at almost any price the purchaser chooses.

What do the names given to wines really mean? Firstly, you may have a genuine wine named after the estate of the proprietor. This you may believe if his brand is on the cask, and his label on the bottle, and if you have imported the wine direct from him, or from a first-class house in the nearest city, say at Beaune, or Bordeaux, or some man of character at home. Secondly, some of the names show the district; for instance, "Fronsac," "Montferrand," &c., come from the *communes* indicated. But, thirdly, with retail dealers the name is often a mere conventional sign of the quality of the wine. Thus, one may see in a wine merchant's list, the names of half a score of villages or towns; Moulin à Vent, 3s.; Savigny, 3s. 6d.; Beaune, 4s.; Pommard, 4s. 6d.; Nuits, 5s.; Voluay, 5s. 6d., &c. &c.; each priced 6d. higher than its predecessor. This merely means that there is a certain standard of goodness, which a wine ought to have when it is called by these names; but if anybody thinks that all the wine ticketed Pommard or Volnay, &c. &c., comes from those places,—why, he may be complimented on his faith.

Now for a few words on the uses of these wines. They are of moderate alcoholic strength, averaging under 20° per cent.; they are perfectly fermented, and free from sugar and other materials likely to undergo imperfect digestion and provoke gout or headache; and they are admirably well adapted for children, for literary persons, and for all whose occupations are chiefly carried on indoors, and which tax the brain more than the muscles.

As for persons whose occupations are carried on in the open air, and require much exertion of muscles and little of brains, there is good beer to be had in abundance, and no better investment of a penny can be conceived than half a pint or a pint of ordinary London porter—call it "cabman's mixture" if you please. But as for the numbers of persons—very poor ones, too—who lead indoor lives, such as teachers, milliners, dressmakers, and needlewomen of all sorts, if they are young, they can drink beer, perhaps, and make up by "antibilious pills" for want of exercise and fresh vegetables. But once past thirty, beer, as a rule, can no longer be taken with impunity by a great many of them; gout and rheumatism take the place of "bilious disorders;" and their choice is between wine and gin. Wine of the best and purest sorts heretofore was virtually inaccessible; now at least it *can* be got by any persons who have the good sense to prefer it to gin, and economy and forethought enough to feel that a saving of a few pence weekly in an habitual article of food is a bad compensation for illness now or hereafter.

I would that my voice could reach the British tradesman. I don't mean the personage who lives out of town and drives into his place of business in a brougham, but the genuine, old-fashioned, portly fellow who stands behind the counter all day, stays indoors all the week, drinks beer at his one o'clock dinner, and gin or brandy-and-water at night; makes up his books on Sunday mornings, takes an hour or so of fresh air between one and three, and then devotes Sunday afternoon and evening to a good dinner, with a bottle of port; has, perhaps, as Charles Lamb said, a bit of sausage with his tea, and a little something warm and comfortable at night. When I look at the enlarging forms of these honest fellows, and think of their food as compared with their work; and further, when I think of the frightful mortality amongst them in cold winters from " bronchitis "—(say, rather, from a blood too thick and a heart too flabby),—I cannot help thinking that if the maid servant were to fetch a bottle of *vin ordinaire* from the cellar, instead of a pot of beer from the public-house, for the family noonday repast, and if it were substituted for the ginnums-and-water at night, our too solid tradesman would have a more useful liver and lights under his ample waistcoat, and would not be nearly so liable to

"Fall as the leaves do, and die in October."

It will be a good day for the morals, health, and intellectual development of the English when every decent person shall on all hospitable occasions be able to produce a bottle of wine and discuss its *flavour*, instead of, as at present, glorying in the *strength* of his potations.

One thing that would go with the greater use of Bordeaux wine would be the custom of drinking it in its proper place *during dinner* as a refreshing and appetizing draught, to entice the languid palate to demand an additional slice of mutton. Physicians who practise amongst town children, of a class in life where prevention is looked to as well as cure, know well the capricious and feeble appetites of many children; how they cut off their fat and the *brown*, and how they reject every morsel at all under-done. Be the case what it may, children *must* have *quantity* and *variety* of food. If not, if the parents content themselves with the slovenly surveillance of servants, who report that Master Johnny is a remarkable child, quite healthy, but wont eat his meat; or that Miss Jeannie is plump, and *so strong* that she takes and requires as great a dose of aperient medicine as a grown man, and that she loves bread and butter and sugar better than meat;—then comes an age—say fourteen to seventeen—when the teeth are found to be decayed, or when the boy or girl is said to have a "delicate chest," and must go to Torquay, or the young lady to some chalybeate water, and all those other horrors too well known to parents of "delicate," *i.e.*, underfed or appetiteless children. Much of this might have been prevented, puncheons of cod-liver oil might be spared at the age of 16–20, if, at the age of 7–10, the governess had said, "Miss Jeannie wont eat her mutton," and if the physician had said, "Give her some kind of light, clean-tasting, sub-acid wine—

Bordeaux or Hungarian—let her sip this, *ad libitum*, at dinner, so that it may tempt her to relish her mutton."

Curious are the social changes of sixty years. Dr. Trotter, who wrote a book on drunkenness at the beginning of this century, denounces the custom of taking wine at dinner. "Thraeum est," he exclaims, "tollite barbarum morem!" To drink *after* dinner was then orthodox. *Now*, we say, drink what you please *at* dinner; the more and the more varied the wine (on festive occasions) the better; but don't *sit* and drink after dinner.

The Bordeaux, like other fine light wines, make pure healthy blood, and at the same time favour the action of the excretory organs; they are good in the anæmia and chlorosis of growing girls. How often I have wished that the patients coming from a Dispensary or out-patients' Hospital room could have had a bottle of such wine, instead of the filthy "mixtures" that they carry away in their dirty bottles! Mixtures, too, contaminated with methylated spirit! which the infernal ingenuity of wholesale chemists supplies at low rates, in the shape of "tinctures," to parsimonious Dispensary committees! O Charity! what crimes are committed in thy name!

To persons of the gouty and rheumatic temperament—maladies which they vainly attempt to keep at bay by the driest of diets, such as meat, bread, and brandy-and-water—Bordeaux wines are of special service; they neither turn sour themselves, nor are they the cause of sourness in other articles of food. But, be it observed, they are *beverages* and not *drams*.

Then what a boon it would be to the very flower of our female population if the medical profession were courageous enough to set at defiance all the army of Mrs. Gamps who infest the lying-in chamber, and who insist on cramming young mothers with the heaviest beer or porter, brandied wine and ardent spirits, on the pretence of keeping up their strength and assisting them to nurse! If ever there were a fit machinery for making women drunkards, it was the whole organization of the lying-in chamber, as it was when I first knew practice, and even that was an improvement on times gone by. A poor woman, after the pains of childbirth, was loaded with bedclothes, and carefully shut out from fresh air and denied wholesome ablutions, in order, as it was said, to keep out the demon Cold. She was starved, denied a slice of roast mutton or any solid food, and saturated with gruel and other fearful slops in order to propitiate the demon Inflammation. Fruit and vegetables were denied, because of the belching demons Acidity and Wind! Then, when duly softened, sweated, blanched, puffy, nerveless, and breathless, she was exhorted to take stout or ale and port wine to keep up her strength and make milk for the little one. How soon young women get a bloated look and lose their youthfulness under this *régime*, every man of observation knows too well. But it is not so well known that in humbler circles, where no port wine is to be had, the gin-bottle was and is the substitute. Talk of Mission Women! Low monthly nurses are the very missionaries of ardent spirits.

But I affirm that, whilst the labouring man's wife, with her active muscular system, can nurse very well on table-beer, and wants not a drop of gin, so the lady, with her more active nervous system and delicate organization, can nurse very well on pure clean claret.

She may drink abundantly of it, and be fresh, young, rosy, and fit for another innings when her duties are over,—with none of the dusky, venous tint of nose and cheeks, none of the misshapen "figure," for which anatomical corsets and belts are prescribed in vain.

One of the chief great medical uses of the Bordeaux wines is to relieve the restlessness, nightly wandering, and thirst of the exanthemata, and especially of scarlatina and measles, in children. If a child is very stout and red-lipped, I should not press the use of wine during the first day or so; neither, in fact, need one *press* it at all. Mix one part of pure Bordeaux wine with one or two of pure cold water, according to the patient's age, and let him drink it at night *ad libitum*. I know of no "diaphoretic," "saline," or "sedative" so admirably adapted to allay the miserable wandering, the headache, and thirst of scarlatina. What an improvement it would be if we were wise enough sometimes to trust to our patient's instinct! It is contrary to all experience that a sick child or other unspoiled person should go on sipping what made its head ache more, or its pulse beat higher, or which added fuel to a tormenting heat and thirst. In measles, so soon as the rash becomes dusky, Bordeaux wine allays the great restlessness. This, be it observed, is not a treatment founded on any hypothesis that alcohol is a good aliment for the nervous system, but on observation of facts at the bedside. It is no more than the small beer which Sydenham used to allow his patients in small-pox and pleurisy.

Any one who observes what takes place within himself may soon distinguish four kinds of thirst. The first is that which arises from want of moisture, as from excessive perspiration in summer, and is almost certainly allayed by water. The next is a false thirst, depending on a disagreeable state of mucous membrane of tongue and fauces. This is common enough with dyspeptic people, and with many children who are "always thirsty," and is not only not relieved but aggravated by copious draughts of cold water. The third is a thirst truly subjective, depending on the nervous system—the thirst of mental agitation, of bodily pain, or of intense fatigue and exhaustion. Any one who has ever experienced this last may know that whilst mere water is only valuable as a kind of diversion, a drop of wine acts magically. When one sees a man, "unaccustomed to public speaking," humming and hawing, and in vain trying to lubricate his tongue with a glass of cold water provided for public lecturers, it is clear that a more advanced knowledge of physiology would have caused that glass to be filled with wine, to oil the brain, which was the really dry place, whereas the jaws might have been left to themselves. A lady complained to me that her daily governess, when she came to her house, always asked for a glass of cold water. It is very common with sickly, bloodless milliners' girls. Fruit, or food with wine, are the true remedies for the foul tongue and nervous exhaustion which the poor creatures delude themselves by calling thirst.

All three kinds of thirst probably exist in the exanthemata, and after the first or second night, if the patient voluntarily sips, and does not reject Bordeaux wine and water, it may be given *ad libitum*. Adult patients have gratefully described to me the extreme refreshment and quietness which such a drink produces throughout scarlatina simplex and anginosa.

Of course, if the patient dislikes it, there is no more to be said.

Fourthly, there is "the thirst that from the soul doth spring"—the craving for intellectual enjoyment and gaiety which wine satisfies effectually and innocently in a way that nothing else can. But let me quote Dr. Guyot* with regard to the effect of light wine in quenching the thirst of the body. He says that, "In England the port and sherry never refresh me; they may be capital *vins de liqueur* for occasional use, but for daily habitual ordinary drink at meals, nothing," he says, "equals the wine of the more temperate regions of France."

I put together three classes of patients—rheumatic, gouty, and bilious—because they are the chief sufferers from heavy, ill-fermented, alcoholized, and ill-blended beers and wines. I have no theories; but state the fact that persons whom I have attended for years enjoy good health whilst they drink pure Bordeaux wine, and suffer in head or joints the moment that they touch port or sherry, unless of the dearest and oldest qualities. Practitioners of the last generation used to be haunted by the demon Acidity, and to think they could cast it out by a diet of meat and brandy. I say, try claret, and you will add ten years to your patient's life and to your own fees.

---

* Sur la Viticulture du Sud-ouest, &c., p. 47, 1863.

CHAPTER VIII.

The praise of Burgundy: its perfume, its place at dinner—A visit to the Côte d'Or—La Fontaine Couverte—"Saintenay, *cinquante-six*"—Volnay, Montrachet—Mersault blanc, and "l'ingrat midshipman"—Clos de Vougeot, Beaune, Beaujolais, Macon, and Hermitage.

O much for Bordeaux wine, on which I love to linger. It is such a model of purity and freshness; so little prone to disagree with any one; so well adapted as a beverage for all ages and all conditions. To me it resembles young, fresh, laughing, innocent girlhood. But there is a something beyond even this. We may admire the rosebud and the snowdrop, but there is a place in our affections for something fuller, warmer, rounder, and more voluptuous. As is Aphrodite to a wood-nymph, so are thy wines, O Burgundy, to those of thy sister Bordeaux!

If any of my readers will do me the honour to be advised to study this wine, let me entreat him not to begin with a cheap sort, but to select a good specimen in which he will find the peculiar excellences well marked. As in studying anatomy the student should get well-developed bones to begin with—then he will detect the various surfaces and processes in less marked specimens—so I advise the student to invest four or five shillings in a bottle of good Volnay. Take care that the wine be not chilled nor the wine-glass cold (58° Fahr.), and drink of it in the middle of dinner with roast meat, or, still better, with hare or other game. One bottle is quite enough for four persons. Then say what you think of it. Is it sweet, or sour, or hot, or strong? No, it is perfume. It is not a liquid *plus* perfume, but it is itself a liquid perfume, one and indivisible. The perfume or bouquet is the first thing tasted and the last. It hangs on the tongue and palate, and leaves a permanently agreeable impression. There is no room, as in some wines we have treated of, for distinguishing various properties after deglutition, even if of perfect unity and pureness. The first and last and only property is the perfume.

And what is this like? It is an intense wine flavour with a trace of bitterness in it. This is not saying much; but if the taste be any guide to its nature or alliances, it ranks with the class of substances of which valerian, civet, and castor are examples. Perhaps, however, in this I am wrong; it may be that the intense *quantity* of wine taste and flavour in the highest Burgundies acts *qualitatively*, as I have shown to be the case with many perfumes.

Burgundy is pre-eminently a full-bodied wine; but its body is aromatic, not alcoholic. Of course, like all great artists, I am drawing from the live model. I write with a bottle before me, which I am sacrificing for my own inspiration and my readers' profit; and the alcoholic strength of the generous liquid is only 22 or thereabouts, whereas a bottle of Cape port sent me by a patient (of course, being undrinkable, it shall be given to the poor) is quite 36. One-eighth of a bottle is as much as a man need drink with the most savoury parts of his dinner; one-fourth of a bottle is quite a good dose for a moderate man. It makes one feel decidedly warmer and more genial; it is a thorough exhilarant, and if taken too freely produces a tightness and uneasiness in the head. But if good, it does not produce any other ill effect; neither does it do so if other wines be taken before and after it, as people ought to do, for to drink Burgundy throughout a dinner is like trumpets throughout a symphony or an apple tart all quince. But if too new, or in a state of fermentescibility or acidity, it will be felt in every joint in the body. *Corruptio optimi, pessima.*

In neglecting Burgundy wine, we ignore a most powerful remedy for poverty of the blood and an ill-nourished state of the nervous system.

A long time ago I learned by experience the value of Burgundy in cases of debility with nervous exhaustion. The patient who first taught me was subject to fits of giddiness or syncope or sudden pallor, followed by hysterical symptoms. He had been subjected to all the artillery of "nervous" and antispasmodic remedies—valerianate and sulphate of zinc, galbanum, &c. His own instincts led him to drink a very good Burgundy, as being more supporting and less heating and acescent than port and sherry, and more full-bodied and satisfying than Bordeaux. I am satisfied that, although out of a million drinkers fewer would find anything possibly disagreeing in Bordeaux than in Burgundy, yet for a large class of people who want support, Burgundy has in it materials which Bordeaux has not. It is more powerful for good, and of course for evil likewise.

To use a rough comparison, Burgundy has fifty times the flavour which port has, or ought to have, with half the alcoholic strength. But this is saying not enough in praise of it: it really soars high above port in the qualities which distinguish wine in contradistinction to spirits. Whoever would add an innocent pleasure to his Christmas festivities, let him hand round a bottle of Volnay or Chambertin with the roast turkey. One of the most eminent physicians in London assured me, that the height of physical enjoyment to him was the moment of sipping the first glass of Burgundy with the second slice of mutton.

But old-fashioned wine merchants will tell you that Burgundy does not answer. "It wont keep," they say, "it wont bear travelling; and it gives the gout." A year or two ago, Mr. S——, an eminent wine merchant, was at Macon at a time when the leading wine-growers of the country had assembled to taste and classify the new wines. He took the opportunity of bringing before them the popular English threefold objection which I have just stated. In answer to the first objection they brought him Burgundy wine a hundred years old, attenuated, but still sound wine. In reply to the second, they brought him some that had been all round the world; and in reply to the accusation that it caused gout, they bid him inquire at dinner-time, amongst more than 200 wine-growers then assembled—men who loved, and drank, and swore by their own wine—how many had gout. He was quite satisfied on this score. My friend L. (who, as a Burgundian, must know) assures me that the only gouty people in the district are those who eat too much, and do not drink enough. And I am sure L. is right.

But it is time to quit these lofty regions, and come back to the cheap wine which is my subject. Within certain limits, the value of wine increases, not as the price, but as the square thereof—*i.e.*, a bottle that can be got for 4*s.* 6*d.*, 4*s.*, or even 3*s.*, is five, four, or three times as good as what can be got for 2s. or 2s. 6d. It is only in first-class wines that you get the totality of flavour which I have been describing; in cheaper ordinary wines you get more or less of it. We ought certainly to get a something fuller, stouter, rounder, and higher-flavoured (on the average, two, three, or four degrees stronger in alcohol) than in the pure, delicate, virginal Bordeaux of equal price. In lower qualities we get a something acid, perhaps very acid, and the flavour is so little, or none, that no one could say what the wine might be.

So far I wrote in 1865, when the former edition of this book was published. Later in that year, so propitious to the vine, I had the privilege of visiting Burgundy and judging of its wines on the spot, as the guest of my friend L., whose family are ancient proprietors and wine merchants at Beaune. I never shall forget the glorious September afternoon when, standing on the old ramparts of Beaune, the whole range of hills rose before us, each bearing some honoured name;—with Pommard and Volnay in front. The next day we drove about twenty miles to the south of Beaune, visiting on our way Volnay, Pommard, Meursault and Montrachet, where I heard the young wine chirping in the casks, like an infant Bacchus in its cradle. We had a long drive through a country *bien accidenté*, passing Chagny and many villages intermixed with vineyards. My friend pointed out the harmony, as it were, of the country—" the Golden Slope,"—the *Côte d'Or* ; how the hills produced wine, the plains wheat, whilst cattle graze on the low ground by the river; and the poplars supply a light wood for packing-cases and double casks. After a time we arrived at a clearing in a wood, where was a square lawn encircled with trees, and hard by an ancient fountain, protected by a brickwork arch, whence the place was called *La Fontaine Couverte*. Here a large party were assembled of the country gentlemen of the neighbourhood and their families for a pic-nic. I should exhaust my whole vocabulary ere I could describe the exquisite grace, the playful beauty of the French girls, the matronly cheerfulness of their seniors, the quiet refined enjoyment of the whole party. We had a capital collation served in the most comfortable fashion. First, there were cold chickens, and secondly, a magnificent pike stuffed and roasted, to be eaten cold; and with them we had some excellent white wine, which was tempered with water from *la Fontaine Couverte*. But now came the climax of the feast; a hare pie, the memory of which it would take long years to efface; and with this, an old family servant came round with a bottle and some glasses smaller than those in which we drank our wine and water. As he poured out each glass, he said with a solemn voice, "*Saintenay cinquante-six. On n'y mêle pas de l'eau.*" The wine, if I mistake not, came from the *vignoble* of a gentleman present. I must not attempt to chronicle the creams and fruit and other good things which followed, including some superb coffee; but justice compels me to add, that when I saw the young ladies each folding a delicate fairy-like cigarette, and saw "the smoke that so gracefully curled," as they lifted it to their lips, I, who never smoked a grain of tobacco in my life, was forced to confess that nothing becomes a pretty woman so much as a cigarette. But I must not linger longer at *la Fontaine Couverte*; let me only record my respectful admiration of a people who clothe the most trivial amusements and social intercourse with such delicate gaiety; and let me once more express my veneration for that veteran servitor who took care that his master's finest wine should be treated with due respect. Such people deserve to have good wine,—and they have it. I shall never forget " *Saintenay cinquante-six.*"

The chief white Burgundy wines are the Montrachet, and Mersault, wines of a splendid straw amber colour, which, without being sweet, have yet great body, suavity, and graciousness. The taste and bouquet it is difficult to describe; I may say they are vinosity itself. In common English parlance, the wines of the adjoining departments of the Yonne, and of the Saône et Loire, including the Maconnais and Beaujolais districts, are reckoned as Burgundies; and in the conventional nomenclature, most of the white Burgundies sold in England, if they are not called Mersault or Montrachet, are called Pouilly or Chablis. Plenty of these white wines are to be had in England under these names; Chablis from 16s. to 40s., Pouilly from 20s. to 50s., and Mersault and Montrachet from 30s. to 70s. Some cheap and excellent wine is sold under the name Mercurey, the name of a village to the south of Beaune. I believe the genuine Chablis is seldom to be had, though a white wine which takes its place is commonly produced at the beginning of dinner with oysters.

In an amusing rhapsody in praise of Burgundy in the *Almanach du Moniteur Vinicole* for 1871-72, the author extols, " ce Beaune première bien connu à la Chambre des Lords! Il combat le spleen et le brouillard." Of the white wine he says, "Ce Mersault blanc console bien des ladys du départ de l'ingrat midshipman! Un doigt de cet or en fusion—un biscuit et une larme—Adieu, *my dear!*" I must express a doubt whether the author is right in believing that Mersault is quite the wine to console " des ladys" for the departure of the ungrateful midshipman. It is not the wine to make a Penelope.

As I said of Bordeaux, so I say of Burgundy, the neglect of white wines is discreditable to the intelligence of the age. The white are more subtle, more excitant, more delicate. They are capable of furnishing innocent pleasure and giving a new fillip to many an invalid. I well remember being consulted by a venerable general on the failure of certain digestive functions, and whilst he was reciting a long and dismal list of drugs, I asked, "Have you tried white wine?" "White wine?" he said, "why, I drink the finest dry sherry that is to be got." But when I suggested to him Pouilly, or Mersault, or Mercurey, it was evident that a new light burst upon him; he never thought of that, and confessed that I had pointed out a most agreeable addition to his small stock of enjoyments.

As for the red wines of Burgundy, I must claim my readers' indulgence in any attempt to describe them in detail. The best are made from very small black grapes, with their skins very austere and sugary, *Le Noirien* or *Pinot*; a second class from the *Passe tout grain*, while the *Gamay* produces a wine more plentiful but less rich. But whilst the vine is one element, the soil,—the site on a hill-side, the *mi-côte* or mid-slope, exempt from the frosts and winds of the hill tops and the fogs of the plain,—and lastly the weather, are of great importance; especially uninterrupted sunshine during the latest stages of the ripening of the grapes, which, as I have said, contain all the elements of grape juice in a concentrated form.

The *Côte d'or* extends from near Macon to near

Dijon, and every village and railway station gives its name to a wine. But the generally acknowledged primacy belongs to a group a few miles to the north of Nuits, including Chambertin, the favourite wine of Napoleon the First; Richebourg, Corton (whose wine is sometimes frozen—*Corton gelé*, see p. 19), Tâche, the favourite wine of the Czar, Romanee Conti, Vosne, Mersigny, and, lastly, the Clos de Vougeot, an enclosed vineyard in honour of which a French colonel once made his men present arms as they passed it, and thus gave a kind of military sanction to its primacy amongst wines. These wines of good vintages cost from 50*s*. per dozen and upwards. Second in rank come the wines of Nuits, Volnay, Pommard; they go at 40*s*. and upwards. The name Beaune is generally given to a highly excellent quality of wine, full-bodied, round, sound, and with abundance of flavour, from 20*s*. to 40*s*., whilst a Petit Beaune, second growths of Mercurey, and a crowd of good ordinary red Burgundies, may be had from 12*s*. upwards. These may be thin, but should not be sour, and at equal price should have more flavour, less astringency, and more body than their congeners of Bordeaux. I asked my friend Laussot to send me in 1865 a case of samples in hierarchical gradation, which he did, and included nine sorts, Petit Beaune, 1859, Beaune, 1862, Pommard, 1862, Volnay, 1859, ditto, 1858, Nuits, 1861, Chambertin, 1850, Romanee, 1858, Clos de Vougeot, 1858, whose prices rose by equal stages from 1f. 25c. at Beaune to 10 francs. There were also Montrachet, 1859, at 3 francs, and Mersault, 1861, at 1f. 50c., miracles of excellence and cheapness. Being sometimes burthened with a multiplicity of samples, I put these by, and forgot them till this September, 1872, when revising this book, and being piqued by the rhapsody about the unfaithful midshipman, I had up a bottle of Mersault. I confess myself perfectly overcome by the excellence of the Mersault, sound, round, buxom, soft, full-flavoured, its flavour fragrant, but with a decided smack of old wine. How people who know that there is such a wine can dose themselves with "natural sherry" is a standing marvel to me. This wine presented a small sediment, of the bundles of acicular crystals, see fig. 1, and rather a large quantity of beads single and in strings.

As I intimated before, I had several opportunities of seeing the process of wine-making in 1865. At Volnay and Montrachet there were small farmers, (each making from twenty to fifty hogsheads per annum,) good homely people. At the Clos de Vougeot everything, from the chief tonnelier downwards, was on a very different scale. A warm drive of about fifteen miles north from Beaune, with the *Côte* on our left, through Nuits, where people were besieging the town well in vain for water, brought us to a large vineyard enclosed in a substantial stone wall, whence the name *Clos*, (*Anglice*, *close*, cathedral close, for instance). In the centre a massive building, half manor-house half castle, contains the apparatus for wine-making, and the huge cellars for maturing and keeping it; the habitable part of the house remains unfinished, chimney-pieces half-sculptured, and walls in the rough, for the place belonged to the monks of Citeaux, whose possession was abruptly cut short by the French Revolution. The four old presses in the press-room are marvels of solidity, and are, I believe, 300 years old. The greatest care is taken that the grapes shall be crushed, and that the contents of the *cuve* or fermenting vat shall be equally mixed; a care that is laudable, though the means employed are not so; because for this purpose men go naked into the vats, to stir up and mix all the grape skins and husks, and, it is said, to promote fermentation by the warmth of their bodies. I understood that a central axis with revolving arms was to be used instead. We saw the wine lately racked off, in barrels most carefully filled, the bung-hole of each covered with a vine leaf; and whenever the *tonnelier* dipped his tube in order to enable L. to taste and to examine the colour by means of that little cup of embossed silver which the Burgundians carry with them for the purpose, then he made a chalk mark, in order that the cask might be filled to the brim at once. There is no good wine in a cask that is not full: this is a Burgundian maxim. M. Pasteur notices the greediness with which the body of this wine absorbs oxygen, and amply vindicates the traditional care taken not to expose it in half empty casks during racking, &c.

In popular language those wines also are reckoned amongst Burgundies which come from the next district southward along the Paris and Lyons Railway. In the department of the Saône and Loire, with Macon for its centre, Moulin à Vents and Thorins, St. Amour and St. Evrient, are wines, some vintages of which are grand, whilst most of them come under the denomination of first-rate ordinary for popular use. The Beaujolais wine is especially to be recommended. I had some admirable samples of Beaujolais from the canton of Belleville, (at 21*s*. per doz. at Lyon,) from the estate of M. le Dr. Burmier, who assured me by a letter in 1867 that he had been fifty years in practice, and during all that time has recommended his own wine to his convalescent patients with unfailing benefit. This was sent me by Mr. A. C. Grove, then resident at Lyons, who also enabled me to judge of some capital Brouilly at 26s. and some Red Hermitage at 31s. and 42s.—wines which are more nutritive and full-bodied than Bordeaux, but do not boast the fulness of flavour of the Burgundies. Richard Chandler, 48, Bread Street, is agent.

It is to the department of Drôme, the ancient Dauphiné, on the left bank of the Rhône, that we are indebted for the famous *Hermitage*, white and red, full-bodied, full-flavoured wines of established reputation. From this same district (*Tain*), Mr. Van Abbott, of Princes Street, Cavendish Square, the purveyor of invalid articles of diet, some time since imported some red and white *Croze* and *Cassis;* wines which, like others of this district as we approach the south, seem to become more full-bodied and less stimulating, rich, round, neither acid nor sweet, and well flavoured.

The department of the Rhone also comprehends the celebrated district known as the *Côte Rotie* and *Vérinay*, whose wines need none of my eulogy.

I do not look upon these as mere matters of commerce, but hold that the man who introduces new articles of food, wine or anything else, for the enjoyment of life, at his own risk and cost, deserves to be ranked amongst public benefactors. If the thing *takes*, the public think he is rewarded, and so he often is; but if the attempt fail, the loss falls on the unlucky wight, who is called a speculator and left in the mire, whilst the British public, like a certain Priest and Levite, walk by on the other side, and reserve their oil and pence for those who are already successful. There is a *Macon Wine Agency* in Old Bond Street, which makes a *specialité* of wines of that district; and there is in Long Acre the establishment of a union of *Wine Growers of Burgundy*, under the presidency of M. le Comte da la Loyère, whose list seems to offer a good choice of the best growths. Their *Fleurie* at 1s. 6d. is excellent, smooth, and rich.

It will give confidence to the clients of this association, and bring a large confluence of purchasers, if I quote from M. Guyot's " Rapport sur la Viticulture de l'Est de la France, en 1863." " M. de la Loyère," he says, " possesses in his estate of Savigny, one of the most ancient vineyards of Burgundy. His labours are incessant and successful in bringing to perfection the best culture and the best viticultural implements; he has created around him a body of devoted and industrious cultivators, whom he rewards with a share of the profits. In truth," says M. Guyot, " when I read on the time-worn stones of the cellars belonging to his ancient feudal château this curious inscription, which dates from the sixteenth century: ' *Les Vins de Savigny sont vins nourrissants, morbifuges, et théologiques,*' I should be tempted to add, ' *bienfaiteurs de l'humanité,*' if this quality were not implicitly comprised in that true and pure theology which alone the wines of France can inspire."

Mr. Rew, of 282, Regent Street, the agent of MM. Piot *frères*, has an excellent choice of Burgundies, " light, rich, and grand," from 24s. to 80s. per dozen, including a Moulin à Vent and Chateau Pouilly, of very fine quality. Messrs. Dymock and Guthrie, of Edinburgh, were polite enough to send me samples of Burgundy which they import, of various degrees of excellence, from a useful " ordinary " at 11s. to a delicious " Chambertin " at 48s. It is a thing of the highest promise for that preacher-ridden people, the so-called Scottish Lowlanders, that a taste for wine should override the craving for *whisky*. Sobriety would then take the place of disputatious Sabbatarianism.

In finishing this meagre account of these fine wines I must add that all authorities, including the Burgundians themselves, admit that there is great room for improvement in the manufacture of many of them; that if the ancient processes of treading and the immersion of naked men of which we read, were superseded, we should hear less of the diseases, the *acescence* and *amertume* to which these superb wines seem beyond many others to be liable.

With Burgundy as with Claret, there is abundance of good, cheap wine, suited to every pocket, which ought to be introduced at dinner and evening parties, instead of vile Hambro' sherry, and to be ordered for the invalid whose appetite needs a whet.

CHAPTER IX.

On some wines of the south of France—Fine vintages—Good ordinary wines of La Gauphine—Béziers—Château neuf du Pape—Roussillon—Lamalgue.

FOLLOWING the division laid down in a former page we now come to the South of France; and again, to clear myself from the charge of presumption I must remind my readers that I am not going to attempt to describe the wines produced by an enormous extent of territory, comprising ancient provinces and kingdoms—Provence, Languedoc, Roussillon, Navarre, Béarn, Gascony—but only to speak of such wines as are get-at-able in England and are of moderate price. Any one who will look at the immense extent of wine-producing departments from the Atlantic, along the north of the Pyrenees, to the Mediterranean, may wonder that their wines are not heard of amongst us. But they are not in their natural condition; for a large part is burnt to make *trois six;* a large quantity is fortified, and much of this passed off under the name of port and sherry; another portion is absorbed by the Bordeaux market, and yet another is used for the *coupage*—*i.e.*, the admixture with the weaker wines of the North, in order to give them the colour and fulness of body which are supposed essential to gain favour with the insular barbarians.

I must be careful not to speak rashly about provinces which possess many vintages of great ex-

cellence, and one of which, L'Hérault, alone yields more wine than the whole kingdom of Portugal; but universal testimony says of *les Vins du Midi*, that, compared with their northern and western rivals, they are coarser, highly coloured, full bodied, destitute of bright distinctive flavour, more often carelessly manufactured, not fermented clean, left too long in the fermenting vats; and *vinés*—*i.e.*, fortified to make them keep, or to convert them into bad substitutes for port. I may say as a matter of interest, that the existence of the evil is evidenced by the number of remedies proposed. M. le Dr. Guyot, in those admirable Reports due to the Government of the Emperor, recommends that the vines should be of better quality, the training and pruning on a different system, and that the vintage should be earlier, whilst the grapes still have a larger amount of acid. M. Batilliat recommends the addition of an extra quantity of tartaric acid to the must.*
M. Maumené recommends the wine makers to mix up the *chapeau* or mass of grape skins, so that the ferment may be kept in longer contact with the must. Suffice it to say, that there is no doubt but that good wine can be made if the makers take the pains, and that if they aim at producing a good stout cheap wine for the English market, they will be abundantly rewarded. M. Guyot quotes abundance of fine wines in the Department of the Gers and Lot, and points to the inexhaustible quantities that may be obtained by the scientific culture of the *Landes*, those vast tracts of seabottom above water which reach along the west coast of France from Arcachon to near the Pyrenees.

---

* Traité sur les Vins de la France, par Batilliat, Pharmacien à Macon, Paris, 1846.

But it is to the Hérault that M. Guyot seems to bid us look for "*les vins de consommation ordinaire,*" and many members of the medical profession are well acquainted with one of the wines which comes from that Department. This is the *Gauphine,* white and red, from the vineyards of Dr. A. D. Du Lac, of La Gauphine, near Béziers. I have seen some dry white Gauphine of nut-brown colour, a most useful wine, which cost about $8\frac{1}{2}d.$ a bottle. Some sweeter white Gauphine, which though sweet was quite sound; and a light amber coloured wine, at about 20s. or 21s. per dozen, which if it could be supplied of equal quality and in large quantities would be a very favourite wine. The *red Gauphine,* is however more popular; a stout, full-bodied, nutritious wine, quite sound; and getting an agreeable flavour if bottled and kept for a year or two. Just the wine for a family of hungry school-girls. The regular transit of this wine was much interfered with by the late lamentable war. A large number of the medical fraternity have used this wine, and I know that one of the officials of the College of Surgeons ordered thirty hogsheads at a time for his own circle of friends. Let me say how advantageous it would be to both parties were the consumer, as M. Guyot suggests, brought into direct communication with the grower,[*] so that the latter might be led by regard for reputation to produce the best wine possible, and the former have the gratification of perfect confidence.

A second batch of South of France wines that has got some repute in a circle of eminent dentists, is that which M. Louis Garcin, of Hyères, sends from a

---

[*] Sur la Viticulture du Sud-ouest, &c., p. 41.

vineyard attached to an ancient chapel dedicated to *Le Père Eternel*. This wine two years ago, of 1868, cost 7 sous the litre at Hyères, and could be laid down in a London cellar by a private importer at 7*s.* 6*d.* to 8*s.* 6*d.* per dozen. When at Hyères in 1870, I had the opportunity of tasting a *Roquette* of 1865, sweet and palatable, without heat, and without any taste of boiled must; also a *Syrats* of 1868, a full-bodied tawny astringent wine; a *Père Eternel* of 1865, of tawny port colour and taste, dry, not hot, nor acid; also a *Carbenet* of 1867; *Briande* 1866; *Fenouillet* 1867, a dry agreeable wine, and a Schieler wine, *gris Gamais*. The wine of *Lamalgue*, near Toulon, is celebrated as a light, dry, agreeable, well-flavoured wine, which can be had of Robert Wood, of New Bond Street, at 24*s.* Some wines from Vaucluse and from Avignon have been brought to London, of these the chief are the *Châteauneuf du Pape*, sold by Rutherford, in Wigmore Street, a very useful stout wine.

I will here merely repeat the notes I made in my last edition about Roussillon. I have not added any since, and I do not want to.

Roussillon is a South of France wine that does duty for port. Of one specimen at 28*s.* per dozen, I note, "Sweetish, on the whole palatable; not hot; yet wanting in vinosity, and having a peculiar earthy flavour of its own."

Roussillon (from a shop in Soho), 20*s.* per dozen. "Acid, terribly rough and coarse; undrinkable." I believe this not to have been brandied. Alcoholic strength $= 23$.

Roussillon (sample from a friend who bottled it a year ago), total cost 20*s.* per dozen, paid 2*s.* 6*d.* duty.

"Coarse, strong, vulgar wine; made me sleepy and headachy."

Roussillon, "sweet and strong; would be liked by women and children; without character and vinosity."

Roussillon, part of the stock of Messrs. S. & R. In quantity can be got for 18*s*. per dozen. Said to be 1858 wine. "Some crust in bottle; cork well tinged, and showing crystals of cream of tartar. First impression, clean, *not hot*. Not too acid. Very astringent; might be called austere. Austerity leaves an unpleasant earthy taste behind it. A little flavour." I believe this to be an honest specimen, and that it would do good service to a purchaser. But alas! purchaser wont buy this as it is: he insists on giving double the price for it, when sweetened and brandied, and converted into public-house port at 4*s*.

CHAPTER X.

Wines of the Moselle and Rhine—"Scharzhoffberger, 1842"—Fictitious Moselle—Excellence of Rhenish wine—First-class growths—Not used as it ought to be—Swiss wine—Yvorne.

Y account of these wines must be very summary, as they contribute very little, and much less than they ought to do, to the habitual consumption of the English, and are rather matters of luxury, medicine, or curiosity.

The Moselle, which derives its origin from the mountains in the ancient province of Lorraine, after receiving the Saar, flows north-east, in a zigzag course, in a deep rock-bound channel through Trêves, to fall into the Rhine at Coblenz. Wherever along its banks (celebrated by Ausonius), there is room and aspect for vines there they are found, and the wine lists of merchants represent a string of villages whose wines are types of a certain standard. Zeltinger, Graach, Dun, Picsport Oberemmel, Brauneberg, Josephshoff, Berncastel, and Scharzhoffberg are the chief names given to a long string of wines priced from 18s. per dozen for Zeltiger, up to 55s. for Piesporter Auslese, or from 6l. per aum of thirty gallons, up to 30l. My petty *cave* contains a few bottles of "1842 Scharzhoffberger," one of which I immolate, and find a remarkably sound, pure wine, fragrant, but not obtrusively so, marvellously clear; its petty sediment free from crystals, and showing a very little torula. No fabricated or mixed wine could have lasted this time, for I am certain of its history during at least twenty-five years. But it must be confessed that the general character of the Moselle still wines, is that they are rather acid, destitute of body, and sometimes artificially flavoured with elder-flower perfume : in fact I have a bottle of this medicament prepared for the purpose. That there are excellent Moselle wines is indisputable ; but they are hardly cheap enough for wide popular use, and perhaps not stout enough. For the rich and luxurious they afford grateful varieties of light fragrant wine, to excite the appetite at the beginning of dinner.

My mention must be short, though respectful, of those noble vintages of the Rhine which afford such models of what wine ought to be. Of light alcoholic strength, and yet almost imperishable through their purity, and with marked fragrance of both the grapy, flowery, fruity, as well as of the true vinous character, the Rhenish wines are the wines for intellectual gaiety. I never drink a glass of good Rhine wine without quoting to myself that sublime description of celestial Wisdom in the Bible, how it passes through all things by reason of its excellent purity (Wisdom, chap. vii., v. 24). They increase appetite, they exhilarate without producing heaviness and languor afterwards, and they purify the blood. But they are not cheap as a rule. In fact they are amongst the most expensive of wines.

So much has been said of the higher growths by travellers, physicians, and dilettanti, that I must cut my description short.

On the right bank of the Rhine, below Mainz, from Eltville to Rüdesheim, a distance of not more than about nine English miles, between the Rhine and the Taunus mountains, lies the Rhinegau, a district of incredible richness as a vineyard. Here are grown Erbacher, Hattenheimer, Steinberger, Rauenthaler, Graefenberger, Marcobrunner, Johannisberger, Geisenheimer; and about four miles lower down Assmannhauser and Bodenthaler. On the left bank, a few miles lower down still, is Baccharach, *Ara Bacchi*, of ancient renown for wine, though perhaps this may be partly owing to its being the wine market for the Rhinegau. All the wines of this district, and from any other of Central Europe, are familiarly called *Hock*, from the vineyard of Hochheim, which produces some of the finest.

I have before me an invoice with the prices paid by a friend who imported some of these wines from an eminent firm at Mayence:— Schloss Johannisberg, 120*s.* per dozen; Steinberg Cabinet, 100*s.*; Marcobrunner Cabinet, 90*s.*; Gräfenberg Cabinet, 83*s.*; Liebfraumilch, 70*s.*; Hochheim, 62*s.*; Rüdesheim, 54*s.*; Geisenheim, 48*s.*; Bodenheim, 36*s.*; Nierstein, 24*s.*; the last three being respectively 25*l.*, 19*l.*, and 15*l.* per aum of 30 gallons.

For ordinary Englishmen of the class for whom I write, to give 10*s.* a bottle for wine is absurd and iniquitous; to hand round such wines to ladies and promiscuous people at dinner (who would infinitely rather have cheap champagne) is a sheer waste. Such wines should not be dribbled out to young women as a foil to fish and *entrées*, but be reserved for the deliberate after-dinner judgment of philosophers; and I may say that the man who is not satisfied with a good Rüdesheimer or Hochheimer at 50*s.* to 70*s.* must be very hard to please. These are the wines drunk by Pallas Athene at the council feasts of the gods. The variety and complex harmony of their body and flavour can only be compared to a chord held down on some full organ. In medicine their uses are incalculable when we want to refresh a vapid tongue, enliven a fatigued nervous system, and yet open all the pores. Examples come thickly upon my memory. A case of almost hopeless bronchitis, patient ready to give up breathing from fatigue and want of sleep; steered her through with Rüdesheimer, and fed her on asparagus. A lady exhausted with the confinement and rest necessary for the repair of a broken limb; flabby, over-stout, and sallowish. In these and the like cases life may be preserved, or a fit of prolonged bad health averted, by the judicious use of these wines.

Most of the "Hocks" are white, but there are some red, as the Assmannhauser. I have had this (Cabinet 1857) at 54s., from Meyers, of Mayence, and found it a wine of great body, powerful and peculiar aroma, and deserving to be alternated with some of the higher Burgundies, to which it is nearly related. Some of it has kept well till 1872, though with a good deal of loose red sediment. Slighter varieties, called Assmannhauser, of lower price, are probably Affenthaler. Some in my possession is very good wine, stands well, and cost 36s. per dozen in 1862.

All the above are dear wines. I am afraid I cannot say much practically about cheap ones. In reading authors of the sixteenth and seventeenth centuries, and romances, such as the "Fortunes of Nigel," one's mouth is made to water fruitlessly by descriptions of "runlets of Rhenish," "flasks of old red Rhenish," which seem to imply that Rhine wine was then in common use here. Better evidence is that of the great physicians and surgeons of the seventeenth and eighteenth centuries, who used to speak of the virtues of " old Rhenish " in relieving wounded soldiers from the effects of scurvy and often of their abominable diet. But it is very certain that no wine is now in less use in England than the cheaper kinds of Rhine wine. The wine merchants' care is to dispose of their "Cabinet wines," and they would sooner get 6*l.* for one dozen of Johannisberger than for six dozen of "Niersteiner." Of course such wines are to be had. Gilbey, Leleux, R. Hall, and other wine merchants supply excellent wine at from 15*s.* to 30*s.*, good wine, but, to the grand sorts, as a whistle pipe to a church organ.

The following memorandum, furnished to me by my cousin, Dr. Charles Mayo, who was Director of the Alice Hospital at Darmstadt during the late war, shows that there is no lack of cheap wine in the country. It came chiefly from Bischofsheim:—

" The wine used in the Alice Hospital between the 20th of October, 1870, and the 1st of April, 1871, amounted to 4633 bottles of white, and 6332 of red Rhenish wine, 60 bottles of Champagne, a few dozens of superior white wine, a few of Bordeaux wine, and about 30 dozen of port. The 10,965 bottles of Rhenish wine were actually consumed at a cost of $17\frac{1}{2}$ kreuzers (a fraction less than 6*d.*) per bottle, but this was owing to exceptionally favourable conditions. It was procured direct from the grower, and the charges for bottling and transport were reduced to a minimum by the kind assistance of Herr Hess, government auditor. The number of patients admitted during the period was 755, of whom 545 had been discharged or 'evacuated,' and 29 had died. The cases were at first chiefly dysentery and typhoid fever, but afterwards there was a considerable proportion of wounded men."

I know that very good Rhine wine could last year be laid down at 1*l.* per dozen by any one who will take the trouble to import it or buy it by the aum. For this purpose neighbours should " co-operate." The squire, the parson, the doctor, the lawyer, the retired officer, and the widow of limited income in a country village, instead of sulkily ordering fabricated sherry by the dozen, should combine and have in and bottle an aum of Hock, a hogshead of red Bordeaux, and one of white Burgundy, and should bottle and divide the product. But I fear that this year (1872) common Hock costs more on the Rhine than in London.

With German wine I must apologize for saying three words about the Swiss. I never got a decent wine at any hotel in Switzerland. Experienced travellers always drink Burgundy. Such as their wine is, the Swiss drink it all, and import more. A very small quantity of white Yvorne comes to England, and is sold by R. Wood, of New Bond Street, a pleasant light subacid wine.

CHAPTER XI.

Wines of Italy—White Capri—Chianti—Montepulciano—Wine at Sorrento—A perfect Hotel—Wine list from Florence—Wines of the Ancients—Vino Rosso Romano.

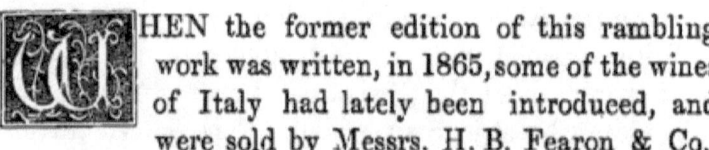HEN the former edition of this rambling work was written, in 1865, some of the wines of Italy had lately been introduced, and were sold by Messrs. H. B. Fearon & Co., of Holborn Hill and New Bond Street. The first I ever tasted was the *White Capri*. Its percentage of proof spirit is 21·8. It is of a light cowslip-colour: fragrant, subacid; brisk, as if slightly aërated; has nothing in it to offend any one; not hot nor yet cold, and seems capitally adapted for young innocent people at their Christmas merrymaking. With such wine at sixteenpence per bottle there cannot be the slightest excuse for poisoning poor unsuspecting boys and girls with Hambro' sherry.

Besides this, in the summer of 1862 was to be had a red *Montepulciano*, from Tuscany, at 14s. per dozen, noted as astringent, clean, light, dry, and wholesome. Also red *Chianti Broglio* at the same price, noted as peculiar tasting, astringent, subacid, and wholesome. Also a red *Barbera*, a Piedmontese wine, at 20s., a peculiarly-flavoured, full-bodied, rough wine; and a *sparkling Vino d'Asti* at 24s., strong, sweetish, unstable, not to be recommended so far as one specimen was a test. I believe that now not all these wines are in the market, but I got fresh specimens of

Chianti and of Barbera, of which this is the description :—

Chianti, 14*s*., peculiarly full coloured, alcoholic strength = 19·8 per cent. of proof spirit; *Barbera*, 20*s*., also peculiarly full coloured; alcoholic strength = 25 per cent. In each great and peculiar astringency. Large quantities are consumed by Italians in London; little by others. The taste might seem unusual and startling at first, yet there is no reason why any one who desires a rough red wine to drink with water might not try these, and might probably relish them.

Since my former edition I have visited Italy, and noticed the series of small books* on wine culture, at Florence, intended to teach the best methods to the wine growers. For, in good truth, the ordinary wine of the country, so far as I had opportunities of tasting it, bore marks of imperfect fermentation and tendency to acidity. A visitor to Sorrento finds it difficult to say whether he admires most the marvellous fertility of the soil, where vegetables, orange trees and vines, on the same ground, drink up every ray of sunshine, or the industry of the people, in cultivating every inch where there is earth enough to give a root-hold. In this part of the country, property seems much subdivided, and wine-making must be on a small scale. The ordinary wine of a good wine-shop, in Sorrento, both red and white, was delicious and fruity; but the way in which it is kept shows the imperfect state of manufacture, and a very ingenious mode of avoiding its consequences. The wine is stored usually not in wooden casks, nor in corked bottles, but in *birettas*, huge glass bottles

---

* Della Vite, dell' uva, e del Vino. Firenze, G. Polizzi e comp., 870.

holding more than two gallons, and protected not by a cork, but by a film of oil which floats on the surface of the wine, and answers the double purpose of keeping off the air, and of allowing the escape of any bubbles of carbonic acid that may arise when the wine takes on a fit of fermentation. It is very difficult to get any drinkable native wine in the hotels — I speak of Naples, Rome, and Florence; the ordinary wines they allow you to have, are simply execrable, and the fine wines not nearly so good as Bordeaux, for which they charge no more. But when I speak discontentedly of the native wine to be had, at the otherwise excellent hotels, in the great Italian cities, I must in common gratitude quote an hotel where everything is perfect, even to the *vin ordinaire* of the place. Any one whoever when ill and weak has been at Sorrento, will know by instinct that I mean the Hotel Tramontano, where the invalid traveller is sure to receive from the English mistress all the personal care and kindness and comforts that he could expect in the most hospitable private mansion.

There are many Associations and individuals anxious to revive the œnological repute of Italy. The slightest acquaintance with classical writers shows that they had wine, good wine, and good old wine. As the distillation of ardent spirits was utterly unknown, they could only have preserved their wines by one of two means—perfect fermentation, or the use of antiseptics, such as pitch or resin. But as these latter means must be intolerable to women and persons of fine taste, we prefer to believe that perfect fermentation was the chief reliance. The *amphora*, or *diota*, a large earthen vessel with two handles (like two ears) near the top, and ending in a narrow point, was well devised and useful for its purpose. As it could not stand upright, it was of course obliged to be supported (it never could be laid on its side, with its large mouth, closed with leather and pitch), and thus all sediment would collect into the narrow bottom.

The ancients seem to have had four kinds of wine. 1st. The ordinary light wine used as the drink for families, servants, soldiers, and the population generally; this doubtless, as it does now, contained a notable proportion of vinegar. 2nd. A wine of higher quality, too austere to be used new—for it passed into a proverb that wine that was pleasant in the cask would never live to be bottled—soon ripe, soon rotten. This wine was matured by being exposed to the sun, or by being placed in properly constructed chambers over the bath; *i.e.*, the Roman or Turkish bath. After this, when well deprived of its lees, it was kept in the amphoræ, which were closed with cork and pitch, "*corticem astrictum pice.*" In these it was kept for some years. A third kind of wine was the sweet, made of dead-ripe grapes; and a fourth, that which was artificially thickened by boiling and concentrating the must.

Sir Edward Barry, in his observations on the Wines of the Ancients, Lond., 1775, says "that we seldom now meet with any good wines imported from Italy. The *Chianti* was formerly much esteemed in England, but entirely lost its character. Large quantities of the red Florence are still imported in flasks, but from the disagreeable roughness and other qualities, seldom drunk. They have a freshness and beautiful deep colour, and are probably chiefly consumed in making artificial claret or *Burgundy* wine, or in giving more lightness and spirits to heavy, vapid port" (p. 442). This shows that the English were then, as they are now, ready to receive wine from any country which will send it to us *pure* and *ripe*.

I have before me the price list of Caminale e Comp., of Florence, which I picked up in 1870, and which contains about sixty wines in cask and seventy in bottle. The price of common table wine, *vino di pasto*, from various parts, was in wood from thirty to sixty francs the hectolitre, or twenty-two gallons, or eleven dozens; of which Capri was priced from 100 to 150; the highest price in wood of the other varieties of Sicilian, Tuscan, Neapolitan, and Piedmontese wines is 220fr. per hectolitre. The wines in bottle were proportionately dearer—the *vino di pasto* from 80 centimes to 1fr. 20c. per bottle; and the next in cheapness were the Asti and other *vini spumanti* at 1fr. 10c. and upwards.

The Falernian, Massic, and other vintages of ancient celebrity, are not named in this wine list. *Lachryma Christi* is a name liberally bestowed on sweet red wines. The genuine is grown on Mount Vesuvius. Sweet white wines are called " Malvasia," or Malmsey.

Mr. Gambardella, of Paradisiello, near Naples, an artist and man of great ingenuity, is endeavouring to find out the vines and viticulture which shall enable him to export fine wines of the antique cast to England.

I received in 1867, through the kindness of Lord Odo Russell, some specimens of " *Vino Rosso Romano* " from the estate of the Marchese P. Pallavicini. I have examined this wine lately, 1872, and find it perfectly sound, bright if carefully decanted from its firm crust, and like a port wine, without sweetness or added alcohol, with good flavour of old wine.

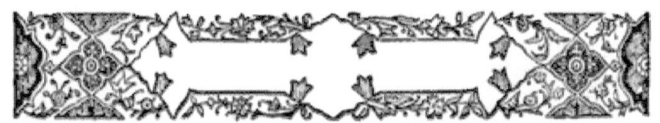

CHAPTER XII.

Wines of Greece—Old St. Elie—White and Red Hymettus and Keffesia—Red Santorin—Thera—Patras—Naussa—Sweet wines—Le Roi des Montagnes.

HE Greek wines, which I only know through Denman, appear of almost perplexing number, and I believe it would be good policy for the vendor to eliminate some of the less important, and fix the public attention on fewer varieties. I first procured these wines in the spring of 1863, and find the following note of *Red Mount Hymettus*, 16s.—" Clean, tasting rough; fair body, not too acid or sweet, something of a resinous flavour—satisfactory." I have since studied them with considerable care, and, to say the least, am convinced that they will form no inconsiderable portion of the future wine of this country, so soon as the middle classes, to whom cheapness is essential, learn to look out for a decided *wine flavour*—that is, for the taste and smell of the grape, more or less modified by fermentation, instead of the taste of spirits. Nay, more, the specimens I have tasted of some of those wines which have had age in bottle, have led me to believe them so capable of developing flavours of peculiar marked character, that they will be sought out for their own intrinsic excellence, cheapness apart.

In order to classify them, we may divide them into

dry and sweet. Of the dry there are the *White Mount Hymettus, White Keffesia, St. Elie,* and *Thera,* the *Red Mount Hymettus, Red Keffesia,* and *Santorin.*

The *St. Elie* at 24s.; alcoholic strength = 25. A light-coloured, firm, dry wine; not too acid; clean and appetizing. An older specimen, which had some age in bottle, was a delicious, firm, well-flavoured wine, admirably adapted for dinner. There seems great promise about it.

So far I wrote in 1865. Subsequent experience enables me to say that the St. Elie went into disfavour with some of my friends from its great acidity and harshness. Blessed is the young wine which has these characters, if only it can be put by to mature. For I find that the St. Elie, if duly allowed to rest, deposits a small quantity of tartar, becomes darker in colour, and acquires a flavour of the true old *winey* character, resembling that of old Madeira. I use the word *winey* to indicate that taste and smell which wine has and which other liquids have not, and which is developed in the intensest form in this wine.

There is another Red wine of a peculiar and first-class character, the Νάυσσα, which is said to come from the neighbourhood of Mount Olympus, in Thessaly, and which is well worth the attention of my readers who want the maximum of vinosity. It is a rich Burgundy-like wine; soon acquires great flavour, but soon also deposits a crust, and becomes of a tawny colour; sound, dry, not acid, and eminently vinous. The Naussa and St. Elie are the monarchs of Greek wines.

There is a peculiarity of flavour about the Greek wines, and there is a trace of it in the before-mentioned *Rosso Romano,* resembling that of Amontillado sherry, and believed to depend on the development of a body called aldehyde, which is alcohol partially oxydized, but not converted into acetic acid.

The White Mount Hymettus, 16s., and White Keffesia, 20s., as I am informed, differ merely in age. Alcoholic strength, about 21. The White Keffesia is a very cheap wine. It has abundance of wine *taste*; whereas some that is older has perfectly astonished me by its firm, dry, clean character, and the abundance of peculiar wine flavour of a Tokay sort which it seems capable of developing.

The *Thera* at 20s.; alcoholic strength = 25. This is a wine which, when new from the cask, is of a darkish sherry colour, full bodied, and very capable of taking the place of "dinner sherry." Comparing this wine with a cheap fictitious wine of equal price, it is instructive to notice the fulness of wine taste and absence of spirit taste. The taste is peculiar; but this wine seems to have great potentiality of developing flavour in bottle. As it is, how superior to cheap sherry! Red *Samos* and *Patras* are new rich wines of full body and peculiar flavour. White Patras is like Chablis.

The red Keffesia, 20s., is a wine of great usefulness now and of great promise, when age shall have matured it. The alcoholic strength is about 21. Full bodied, dry, markedly astringent, not acid, they are much more satisfying than pure Bordeaux of equal price.

The Santorin at 20s. is a very useful wine; it has the tawny colour and dryness of light port, with alcoholic strength about 24 or 25. I have occasionally given or prescribed this wine to poor patients, and been quite amused at their approbation of it; so like port! A dyspeptic, overworked dispensary medical officer, to whom I gave some, tells me that it suits him to a nicety, and controls the acidity of the stomach.

I have had one or two old samples of Santorin of great merit, as being reproductions of certain characters of old port. Again, I have tasted some too thin, slightly acid, saltish, and depositing its crust more rapidly than such a wine ought to do.

The *Como* of 1861 is a wine which seems to have been artificially fortified to imitate port. It is one of the false steps which wine growers took a few years ago, when, instead of trusting to the excellence of their produce, and having faith that anything good in itself will be liked, though its flavour may be new or *sui generis*, they tried to imitate port or sherry. This Como is a good imitation of new port. Price 30*s*.

Como of 1862, price 28*s*.; said to be natural unfortified wine, is extraordinary stuff, and deserves the attention of hospital and dispensary committees. Its alcoholic strength is 30 (it paid shilling duty only); its specific gravity, 1020; and it is intensely sweet, full-bodied, rough, and grapy. Some specimens have a smack of boiled must.

Comparing Greek wines with Bordeaux of equal price, there is more *body* in them, using the word body to imply fulness and rotundity of taste, and what satisfies the stomach, apart from flavour and alcoholic strength also. Persons who might think Bordeaux thin and sour might be satisfied with red Keffesia; on the other hand, a person who delights in light Bordeaux might think the Hymettus coarse, unless he got some of the older and more mature kind. Wine flavour, I need scarcely repeat, is a product of time, and time adds greatly to the cost of wine, so that in cheap wine we regard not so much present flavour as firmness and soundness, and capacity of keeping till flavour shall be generated.

It follows that the persons to whom we should recommend Greek wines especially are those who are hardly weaned from brandied wine, and who require something full-bodied. I find the Red Hymettus much relished by a patient in an advanced stage of phthisis, who says he really prefers port, but that it makes him too hot and thirsty, whereas the Hymettus quenches his thirst, and gives him "support" besides. A second patient, who has had a narrow escape from puerperal fever, says it agrees well, and has checked diarrhœa. The former patient can afford what he likes; the latter, if she had not the Greek wine, would have been condemned to South African port.

Of the sweet wines I have not much to say, as they are not a class of wine that suits me. The Visanto is a very full-flavoured wine, of very high specific gravity and little alcoholic strength. It is found to be good in tipsy cakes and puddings. The Cyprus, Malvoisie, or Malmsey, is marvellously high-flavoured and sweet, and other wines, as the Lachryma Christi, Calliste, &c., partake of these qualities in a lesser degree. Can a patient digest sugar, and does he require it? If so, these wines, with cake or bread, would make a good light refection. They would suit bridesmaids; possibly nursing mothers, children recovering from illness, &c. I have had one bottle of old Thera, marvellously soft and fine-flavoured, though a little too luscious for me.

Sir Edward Barry (1775) says of the Greek wines at that day, "that as from their peculiar fine flavour they are not easily adulterated, they are seldom imported." The best of them come from Santorin, the volcanic isle in the Archipelago, whose wines long ago were stamped with the approbation of Edmond About, in *Le Roi des Montagnes*.

CHAPTER XIII.

Hungarian wine; well known in the seventeenth and eighteenth centuries in England; praised by F. Hoffmann—A visit to Pesth—The Moslem wave beaten back—Tokay—A digression on sweet wines in general; sweet Ruszte, Cyprus, Malmsey, grape syrup, Constantia, Lunel, Rivas Altes—Syrian wine from Mount Lebanon—Australian Cyprus.

I SHALL never forget my surprise and pleasure at first tasting, in 1863, some Ofner. It seemed as if a new field of wine were opened for the English physician. Since that time, Hungarian wine, chiefly through the enterprise of Max Greger, has become familiar as a household word, and has its reputation established. But it is amusing to look back and see how different things were ten years ago, and what a fury was excited by the Hungarian wines amongst orthodox old wine merchants, who with a teleological faith held that the English public were created to drink respectable port and sherry, and that all other wines are "fancy wines." I recollect an argument with a gentleman of this persuasion who tried to demolish Hungarian wines, by asserting, first, that Hungary was a damned out-of-the-way place; secondly, that Hungarian wines were "all muck;" thirdly, that there were no Hungarian wines, and that what were so called were fine French wines sold at half their value.

But it is some vindication for this generation in its taste for these wines, to find that there is not much that is new under the sun. Our forefathers knew Hungarian wine very well, although their name has since been blotted out for a time by port, war, and prejudice. Sir E. Barry (1775) quotes from a paper on Hungarian Wine, by Mr. Douglas (*Phil. Trans.* vol. lxiii. p. 300,) that "the Buda wine is very like Burgundy, and perhaps equal to it. A German author of the last century (Hevelius) says that a great quantity of this wine used to be sent to England in the reign of James the First overland by Breslau and Hamburgh, and that it was the favourite wine of the court, and all over the kingdom." "I remember," says Sir E. Barry, "that several years ago a report prevailed that the Empress-Queen (Maria Theresa) from a grateful sense of the obligations she owed to this kingdom, had proposed to open a commerce from the port of Trieste for more easily supplying this country with several of the Hungarian wines, as the land carriage, which was the most expensive article, would be much lessened by the situation of many of them being at no great distance from the Danube, in which they may be transported a considerable part of the way to Trieste, though with some interruption of the land carriage. But as we were then supplied with good wines from France, Spain, and Portugal, and as the wine merchants did not choose to alter the course of the wine trade, this proposal was no further pursued. But it is some pleasure to reflect that by this channel in some other more convenient time, we may be supplied with a variety of strong and light wines, in their native purity and at a moderate price."*

Thus, with a prophetic voice, does Barry speak, in 1775. But a century before that, Friedrich Hoff-

---

* Barry on Wines, &c., Lond. 1775, p. 466, &c.

mann (*b*. 1660, *d*. 1742), the professor of medicine at Halle, the friend of Boyle, the source whence Cullen, Hunter, Abernethy, and the other lights of physic down to our own day, have derived their inspiration, a man of genial and conservative principles, devoted an Inaugural Thesis, 1685, to a dissertation *De Vini Hungarici excellente naturâ, virtute et usu*.* He begins by ridiculing the notion that the physician should confine his attention to drugs, and not rather to regimen and diet, of which last, wine forms a conspicuous part. He gives five marks whereby the Hungarian excels other wines, and in this he chiefly refers to the sweeter wines of the Tokay order. They are strong, preserve their sweetness, have spirit, odour, and aroma; are strengthening, and yet open the pores of the skin and other organs, so that they cause no headache nor languor; and that the better wines keep for unlimited years. "No country in Europe produces a greater quantity and variety of excellent wines than Hungary," said Mr. Douglas.

I believe the first person who imported Hungarian wines in this generation was the late Mr. Richard Garrett, the eminent manufacturer of agricultural implements, in Suffolk, whose firm has an establishment at Pesth, and who told me that the wine which he imported thence had given him great relief from a debilitating malady which he had suffered from.

I had the opportunity, in 1865, of paying a visit to Pesth, and of enjoying kindness and hospitality there which I shall never forget. Our route lay through the Jura, Neufchatel, and Zurich, but it was not till we reached the hospitable hotel of Das Hecht, at St.

---

† Op. Omnia, tom. v. p. 356, fol. Genevæ, 1740.

Gallen, that we found Hungarian wine in the wine lists. An admirable bottle of old red Hungarian was served at the one o'clock table d'hôte, a fitting prelude to a visit to that cloister and library, rich in monuments of ecclesiastical antiquity, and once the residence of that monk who wrote the imperishable Sequence, "In the midst of life we are in death," which is ensbrined in the English Prayer Book.* There was no lack of Hungarian wine at Vienna. As for Pesth, Buda, or Ofen, whichever name we choose to adopt, the hills on the south of the town covered with vines, the Government Institute for the experimental trial of the different quality of grapes, the hot sulphurous springs at their base—whence the name Ofen, the flourishing firm of Iälics and Co., where I first saw wines exposed to the practice of heating, after M. Pasteur's method : all these things gave the promise of plenty of good wine, so far as natural advautages and human culture can ensure it. The tourist who sees the first outpost of Moslem civilization at Ofen, in the shape of the tomb of a sheik, to which year by year people still come from Stamboul to offer prayers, will be conscious how the followers of the Prophet, two centuries ago, threatened the heart of Christian Europe, and how the stalwart borderers of the Danube, in preserving our faith, also preserved our wine.

Let us suppose that the conscientious student has ordered specimens. Probably a laudable curiosity, combined with the wish to do homage to the female members of his family, will tempt him to begin with a bottle of Tokay. This is by no means a cheap

---

* See *Cantarium Sti. Galli*, St. Gallen, 1845.

wine, for it costs from 3*l.* to 6*l.* per dozen, for pint bottles; but besides its reputation, it is worth studying as a kind of landmark or standard. We need not repeat the information to be found in every book, that it is made of the *essence*, the juice which flows spontaneously without pressing from the finest over-ripe grapes. The result, as it reaches us, is a wine of delicate pale tint, in which the sweetness and fragrance of the grape, though perceptible, are partly hidden by, or converted by age into, an exceedingly rich, aromatic, mouth-filling wine-flavour, so that, rich as it may be, it is not cloying nor sickly, and in its admirable aroma there is a decided remembrance of green tea. M. Diosy's is superlative.

Of course, Tokay can hardly figure in a list of cheap wines; yet it is really cheaper than it seems, for a very small quantity suffices. There are, as I am informed, large quantities in the English and French markets, which meet with a slow sale (at least here), because there seems no place for it in ordinary society. English customs are more and more adverse to sweet cakes and wines, for morning callers, &c.; and English meat-eating people prefer a dry kind of wine. Yet I conceive not only that this wine may be useful as a cordial for the aged, but one bit of experience shows in what respects it is preferable to some other cordials. A short time since I was attending a gentleman, nearly eighty, dying with senile decay and atrophy of the heart, with probably some obstruction to the circulation through the lungs, for though the air entered forcibly, the dyspnœa was most intense—so intense that the act of swallowing could only be performed by snatches, and every movement and everything that " caught his breath " threatened instant suffocation. Having worn out every form of nourishment and stimulant I could think of, at last I suggested some Tokay, which the patient eagerly caught at, and a servant was despatched to get a bottle. The wine merchant had none; but, as it was late in the day, very properly sent on trial what came nearest in his opinion—viz., a bottle of very fine old Malmsey. Next day I had the opportunity of judging of both wines, and of their adaptation to the case in question. The Malmsey was uncommonly fine, rich, and old, but, though mild and soft, was very strong; the alcoholic potency was unmistakeable, and it caused distress to the patient, who could not drink it undiluted. The Tokay from Max Greger's, on the contrary, was marvellously fuller flavoured, and had no prominent alcoholic character at all. It was curious to notice how superior its true wine-body and flavour were to the less winey and more spirituous character of the Malmsey; and the patient swallowed it easily. This may give a useful hint to some of us who are at our wit's end with a patient ill of diphtheria, hopeless phthisis with aphthous tongue and throat, &c., &c. In this case I also ordered a mixture of Tokay and cream. Such things may sometimes soothe a dying bed, and enable an old man to forget the peevishness of suffering, and to bless his family tranquilly before he falls into his last sleep.

It is a violation of logical order, for I intended to have had a chapter exclusively devoted to sweet wines, but as Tokay is the prince of such wines it will save repetition if I sum up what I have to say on them in this place.

There are many wines that are sweet, such for

instance as "Spanish port," and "St. Pancras sherry," but in the list of genuine "sweet wines," I include those only whose sweetness is the result of concentrated grape juice, well fermented, and without the addition of spirit. Of such wines, the Bordeaux district gives the Château Yquem; the south of France offers a great variety, as the Lunel and Rivas Altes; Constantia is a unique example from the Cape of Good Hope. Moreover such wines may be red or white, and of the red the true Lachryma Christi, whether that which is supposed to be prepared on the slopes of Vesuvius, or by M. Gambardella on the other side of the Bay of Naples, and the Greek Lachryma Christi. Such wines ought to be the outcome of the tears or essence of the ripest grapes, allowed by slow fermentation to develop the subtlest yet blandest, most clean and frank flavour. Next in estimation to the essence wines, says Hoffmann (op. cit. vol. v. p. 357), are the *Ausbruch* wines, in which more pressure has been used; and in this category we must put Max Greger's "*White Rustze Ausbruch, No. 18*," and the "*Red Menescher Ausbruch, No. 21.*" All these wines ought to be grapy sweetness itself, without the heat of added alcohol. Another class of sweet wines have a taste totally different; they taste of sun-dried raisins, or of boiled must, and good specimens are very delicious (*vini cotti, vins cuits*). Such are the real Cyprus, the Malmsey or Malvoisie, the Tent, the Greek Visanto. Of these the Cyprus is remarkable for the amplitude of its flavour, so is the Visanto. The practice of boiling down grape juice, or must, into a thick syrup or semi-solid mass, is one of the most ancient modes of making a vegetable preserve. The Greeks called it ἕψημα and σιραῖον, the Latins *defrutum* and *sapa*, the Hebrews *debash*, and the Syrians *dibs*. This has been used from time immemorial as a means of increasing the saccharine element in wine. It is that to which brown sherry owes its sweetness and colour, and is tasted in the Tent wine which is used in English parishes for the Eucharist. Rembertus Dodonæus in his Historia Vitis Vinique (*Coloniæ*, 1580, p. 17,) says that the Belgian and Dutch wine merchants made mixtures of boiled must and Spanish wine, which they sell for Cretan and Malvoisie. There are not many things new.

We thus see the abundance of sweet wines, and in the next place may note their former popularity amongst physicians. "Italiæ vina et generositate et dulcedine superare alia," says Kaupper,* in 1703. Sweet wines are still esteemed on the Continent for their recuperative powers, where Malaga holds the place in public esteem that port used to hold here. Sugar also is far more highly thought of as a restorative; see, for instance, Brillat Savarin's well-known *elixir vitæ*, made of chicken soup and sugar candy. On the other hand, it is the custom with Englishmen and English physicians rather to cast a slur on sugar and on wines containing it.

Doubtless the test of perfection in wine, as a rule, is the fermentation of all the sugar, dry. And we may say that a sweet wine to be wholesome should be free from ferment, be quite clean, and capable of self-preservation without the addition of alcohol. The unwholesome and gouty wines *par excellence* are the sweetish wines with suppressed fermentation.

But with this proviso, we need not forego the

---

* De natura et præstantia Vini Rhenani, 1703.

reasonable use of sugar in wine or out of it. Of the importance of sugar there is no doubt; all starch is converted into it in digestion. Liebig's food for infants consists really of starch so converted. The liver is a great sugar-making organ. We in England are spoiled children; we forget that *honey* was the sugar of the ancients, and that the *saccharum* or *zuccharum* was a medicament, or curiosity or delicacy, only to be met with in the apothecaries' shops, down to the end of the seventeenth century. We look on it as a drug, the parent of acidity, and despise it. Other nations find it more congenial to their digestion, and a force-creating nutriment. Experience, anyhow, cannot be wrong: not Crœsus himself would give half-a-guinea a pint for Tokay, and drink it regularly, unless he felt some benefit from it; but this is what rich old men do. It is often the physician's duty to prescribe about seven meals or refreshments in the course of a day for debilitated patients; when rum and milk, and Devonshire cream, and chocolate, and jelly, and cod liver oil, and arrowroot and isinglass are worn out, then the little glass of Tokay, or sweet Ruszter, or Constantia, or Visanto may come in to lessen the task of the stomach by giving it a little variety.

As this chapter is already out of order it will be no additional offence if I speak of one wine more, to show the growing interest of the whole universal civilized world in wine culture. In 1868, my much-esteemed friend Signor Giuseppe Churi, the Maronite, who once acted as dragoman to the late Captain Sir W. Peel in his travels in Africa, and is now settled at Beirout, sent me specimens of wine grown on the slopes of Mount Lebanon. The wine is of four sorts, labelled—

1. "Vin doux Rosa du Mont Liban, 1864, d'Assad Ebin el Moghobeï;" 2. "Vin doux Rosa de l'Antiliban, 1865;" 3. "Vin blanc des côteaux du Temple de Balbek, 1863;" 4. "Vin d'Or sec du Mont Liban, 1863." The get-up of the samples is most creditable to the good taste of the grower, Assad Ebin el Moghobeï, for the bottles are beautifully packed in rush basket-work and nicely labelled. The *vin doux Rosa* is of a pale pinkish hue, sweetish, but quite stable, and with a decided fragrant taste. The other wines also have considerable merit, and could compete with the dry Ruszte, but are not dry. Nevertheless they have quite the potentiality of being developed into good, full-flavoured white wines under the care and skill of the grower. Anyhow, there they are, good and sound in 1872. Any philosophical œnologist of cosmopolitan tendencies who desires to complete his collection of wines from all parts of the world will be glad to hear that these wines are procurable from Signor Churi at no great cost. Besides, people who are interested in the Holy Places might like to have these wines for sacraments, weddings, etc., just as they do the waters of Jordan at baptism.

Lastly, Australia promises to send sweet wine, if we will take it. I have had from the Auldana Vineyard some samples of sweet wine of a light reddish brown colour, and with the peculiar and unmistakeable perfume of Cyprus.

CHAPTER XIV.

Hungarian Wines continued—Dry White Ruszte, Szamorodny—Dioszeger Bakator—Œdenburg—Somlo—Neszmely—Badasconyer—Red Ofner, Szegzard, Erlaure, Carlovitz, Visonta,—Gold label Ofner—Transylvanian wine.

TOKAY is a classical wine, and its name tempts one; not so many others, whose names would have puzzled Milton's "Stall-readers," and recall his sarcasm on the names of some of those northern preachers who attacked the Greek title of "the book called Tetrachordon," which he wrote in favour of divorce.

> "Cries the stall-reader, 'Bless us what a word on
> A title-page is this!' and some in file
> Stand spelling false, whilst one might walk to Mile-
> End-green. 'Why is it harder, sirs, than Gordon,
> Colkitto, or Macdonnell, or Galasp?'
> Those rugged names to our like mouths grow sleek,
> That would have made Quintilian stare and gasp."

I may plead that I have got them up pretty accurately, and find, besides, that such words as Szamorodny and Dioszeger Bakator Auslese are, thanks to the wine they designate, by no means too rugged for ladies' lips.

Turning, then, from the sweet to the dry white wines, if we take them in a descending series, I will mention first Max Greger's *Ruszte Ausbruch, finest, dry*, at 54s.

This *dry Ruszte* is a remarkably fine wine, and, with peculiarities of its own, resembles some samples I have tasted of first-class white Burgundy, or of a dry St. Peray. Writing with some of these Hungarian white wines on the table before me, it is impossible not to be struck with their admirable fragrance, and how they bring before one the vision of flowers, and likewise of *honey:* not the sweetness which is common to all honey, but the fragrance which is peculiar to the best, and which seems to be of the same nature as this grape perfume. This is especially noticeable in Tokay, and it was noticed by a veteran wine-taster to whom I gave some old choice White Mount Hymettus, which he said had a decided Tokay flavour. Any one who has ever tasted old dry *mead* knows it, but I shall descant on that old English wine presently. Now if I taste, side by side, a bottle of the Szamorodny, described as " a dry Tokay wine," I find in said Szamorodny a most agreeable sound wine, with prodigious fragrance, great dryness, and fine wine flavour; a mouthful is a nosegay. But a sip of the *dry Ruszte* puts out the Szamorodny; there is in it more bodyfulness and a *souvenir* of Burgundiacal bitterness, which shows much greater potency and value. I affirm judicially that the *dry Ruszte* is a great acquisition, and that no dinner will fail to be gratefully remembered at which a man of sense first becomes acquainted with it.

The " *Szamorodny,* or *Dry Tokay* " of Denman, at 42*s.*, is a very fine specimen; and " *Szamorodny Muscat,* first quality," of Max Greger, at 48*s.*, has a prominent flavour of the muscat grape in addition. I need hardly say how the flavour of all muscat grapes is preserved in the wine that is made from them. I have not taken the alcoholic strength of these wines, nor is it necessary. Colour, flavour, and the whole purity and fineness of the wines, forbid the suspicion of such a suicidal sophistication as addition of spirit. Besides, I would rather drink the wine than distil it.

Taking the Dry Ruszte and the Szamorodny as examples of wine possessing, as it were, a duplex aroma (*i.e.*, the fragrant grapy and the true vinous), we may take the *Dioszeger Bakator* as an example of a wine with single aroma of the fresh, fragrant, grapy order. I have already mentioned one specimen of this wine which I got from Denman's, price 32*s.*, and which particularly pleased me. My note was "very agreeable, clean, grapy, fruity." Another specimen from Max Greger's, "Count Stubenberg's own growth," price 36*s.*, is on the table before me, and deserves at least equal praise.

The *Œdenburg* (No. 16 on Max Greger's list) is a lower-priced dry white wine, of the fresh, fragrant, flowery, unaltered grapy flavour. *Delicacy* is the charm of wines of this class; and if they have a fault it is that of being a trifle too thin, so that the acidity is not veiled. Both of these are delicate, one with slight muscatel flavour, and body enough. Alcoholic strength about 20. They compete well with high-class Rhine wines.

The *Somlau*, No. 10 on Max Greger's list, price 26*s.*, seems a sound, dry, firm wine, with plenty of flavour, and cheap for the money. Alcoholic strength about 21.

The *Neszmély* (No. 8 on Max Greger's list, price 18*s.*) is a very cheap wine of slightly darker colour, not deficient in flavour of the vinous order, seeming as if it were a diminished example of the Szamorodny.

The *Badasconyer* I noted as alcoholic strength about 21; aroma full and peculiar, and of the vinous order.

White Tetenyi, 18*s.*, from M. Diosy. A peculiarly nice grapy, well-flavoured, appetizing wine. Very cheap indeed. It is said to be sometimes put into little brown flasks with handles, and sold at a high price as Stein Wein.

Bakator Muscat at 32*s.*, and *White Diasi* at 36*s.*, from M. Diosy, are grapy, muscat-flavoured wines.

Somlo', from the Bishop of Veszprém; price 60*s.*, from M. Diosy, is a very delicate, fine-flavoured, first-class wine; well deserving the attention of the connoisseurs who do not hesitate to give from 5*s.* to 10*s.* per bottle for "hock" for the earlier part of their dinners. This wine must be preserved from sudden fluctuations of temperature.

I learn that the word Szamorodny signifies "self-born," indicative of the spontaneity, vigour, and naturalness of wine.

If rational people must economize in the wine they drink at dinner, why martyrize themselves on bad sherry when such cheap and fragrant wines as these are to be had so easily?

Here let us pause one moment, because that knowledge of wine which every medical practitioner ought to possess, and which seems difficult at first, owing to the infinity of names and qualities, may be rendered easier of attainment by any glimpse, however partial, of a reasonable classification.

White wines, then, in which we decidedly taste the grape, form, as it were, a sub-kingdom of themselves, and it is interesting to observe how the actual state of the grape may be tasted in many of them. Thus there are many which speak out for themselves, so to say, and bear testimony to the fact that they are made of grapes which have not attained the fullest maturity of sweetness,—whether from a northern climate, as many of the Rhine and Moselle wines, or from a cold northern aspect, or from the gathering of the grapes at an early stage. Hence a kind of greenness, as it were; a very light straw, passing into a greenish colour, and general characters of grace, juvenility, delicacy, insubstantiality,—just the characters of a young girl with a young head on her shoulders. Such wines, if of bad quality, are thin, poor, and sour, as aforesaid; if of good quality, their very acidity tends to the generation, in time, of the most exquisite superadded bouquet. Such a wine of low quality is the first met with by the tourist who proceeds along the western coast of Brittany to the south. At a village named Sarzeau, famous as being the birthplace of Le Sage, they make a wine of the small white grapes, which is put on the table at the inn for the guests to drink of gratis, as they do cider elsewhere in Brittany, and water in England. It is sold at five sous per bottle. Similar wine, of rather more stoutness, is made near Nantes, and sold at ten sous per bottle. Such is the wine to be expected at the northern limit of vine culture, but the Hungarian Dioszeger Bakator and Œdenburg are good specimens of the more generous wines of this sort, so are two wines that I have not mentioned yet, the *Pesther Steinbruch*, at 26s., and the *Villanyi Muscat*, at 24s. Each of these is about alcoholic strength 21, and there seems little difference between them.

I have had samples of a remarkably good selection of white Hungarian wines from Hudson and Collins, of Victoria Street, beginning with a wonderfully cheap white Erlaure at 17s., and ascending the gamut of flavour through Œdenburg and Villanyi Muscat, up to Szamorodny and dry Ruszte.

Can we, then, venture to guess at the conditions under which these delicate wines would be appreciated and useful? I think so, if we remember what a help the fresh acid fragrant juice of the lemon is to a man who is eating something too sweet or too rich; and how the Orientals squeeze the half-ripe grapes to make delicious sherbet, and sauce for their *kibobs* and *pilafs*. These are wines for delicate refined people: it would be of no more use to give them to a day labourer than it would be to use a lancet to chop sticks. They are wines better adapted for hot weather than for cold. If a man dines on a single joint, he would prefer a bottle of Erlaure or Ofner; if he has a complex repast, he would drink these light wines with his fish or *entrées*. Give one of these wines to a man whose tongue is too red, and who has diarrhœa, and he would reject it; on the other hand, a man with a clammy-coated thirsty tongue would probably drink them greedily. I may add that I am satisfied of the entire wholesomeness of these white wines. I have drunk of them freely, both singly and in combination, and one night when very tired took half a bottle of the dry Ruszte with unmixed pleasure at the time, and slept soundly and awoke as if I had had "food for the nervous system." It is worth knowing, too, so far as my limited experience goes, that they are slightly aperient, or at least the reverse of constipating.

So much for the white Hungarian wines. But, as with all wines, so with this: the white is more appreciated as a luxury or medicine; the red is that which is most extensively used, and there can be no doubt, from the universal testimony of invalids, that these wines, and especially the Ofner and Carlowitz, are possessed of highly nutritive virtues. I give the following as an unabridged version of my original verdict on these wines :—

One of the commonest and best kinds of red Hungarian wine is that which is called *Ofner*, from the town of Ofen (Buda, Pesth), near which it is made, and of which some varieties have distinctive names, such as *Adlerberg, Blocksberg, Burgerberg*, &c., from various hills in the neighbourhood. My first essay of these was of a specimen of *Ofner*, price 24*s.*, from Denman, in 1863, of which my note is " apparently pure and full-bodied; not acid, nor astringent, *quere* sweetish, agreeable, and satisfactory." Specimens from the same dealer, at the same price, in 1864, sp. gr. ·995, alcoholic strength about 21, deserve the same note, except that I should substitute the words "fruity" or "grapy" and "smooth" for "sweetish." There is also a fine *Ofner Auslese*, at 36*s.*, No. 31 of Max Greger's list, which is very good indeed, pure, smooth, and delicate. I should be inclined to recommend a good Ofner, as I would a good Bordeaux, to any patient whose veins wanted filling with good blood. Some Ofner there is which reminds me, with a fine frank flavour, of green tea and fresh seaweed.

Next I may take the *Erlaure*. A specimen of Max Greger's (No. 27), at 17*s.*, was some time ago brought to me as a rarity, and I was led to believe it was an Assmanshauser. I found it a pure wine, with something of an old dry quality, subaustere, subacid; no volatile bouquet, but a pleasant vinous taste, greatly enhanced by nursing it up to 60°, at which temperature its slight austerity and acidity greatly decrease.

I could find in it no high vinous character, and was greatly relieved when told it was a seventeen shilling wine, at which price it is uncommonly cheap and satisfactory. An *Erlaure*, from Azémar, at 24s., did not seem to belong to the same family; much fuller and fruitier; specific gravity ·995; alcoholic strength about 22. A good useful wine, but I could not have distinguished it from a Szekszard. An *Erlaure*, at 30s., of Denman's, has received great commendation at my table from friends, who have pronounced it " an excellent claret," a verdict I concur in.

A *Visontá selected*, No. 4 of Max Greger, price 24s., alcoholic strength about 19, was noted as a wine of good colour, body, and flavour; quite satisfactory.

I once fancied that I could tell a *Szekszard*, of which I have had some from Denman at 16s., from Azémar at 18s., do., No. 3, from Max Greger at 20s., which I noted as powerful-tasting and strong-bodied, though its alcoholic strength was only 19. "The *Sexard* wine," said Mr. Douglas, in 1763, " is strong and deep coloured, not unlike that wine of Languedoc which is said to be sold at Bordeaux for claret. The Sexard wine on the spot costs only about five cruitzers, or twopence-halfpenny a bottle. It belongs to the Abbot of Custance, and is chiefly consumed in Germany. Sexard is on the Danube between Buda and Esset." The *Menes* seems a full-bodied wine, insomuch that I have suspected some specimens to have been a *little* fortified, but perhaps I am wrong.

Of the *Carlowitz* I have drank an ordinary sort of Max Greger's No. 6, at 24s., and a " selected " of Max Greger's No. 25, at 32s., specific gravity ·996, alcoholic strength 21, full-bodied, subastringent wine of *drier* character, with much of the ruby colour and astringent smack of *port wine*. M. Diosy has a *Carloviczi*, red, grapy, and rich.

Red Tetenyi at 18*s*. I did not like this. It had a perfume of raspberries, and seemed too thin and acid.

Red Diasi, price 32*s*., bottles included, from M. Diosy. Very cheap, satisfactory, soft, smooth, grapy wine, apparently nourishing. Alcoholic strength about 21·5.

Red Visontai, 1854, at 36*s*., from M. Diosy. This, like most other of the best wines, is said to be from the cellar of the Convent. Alcoholic strength about 27. This is a wine of striking character, dry, sub-austere, of potent vinosity; might pass for very light dry port.

Dealers in French wines are legion, and are well known and established; the Hungarian, Greek, and Austrian dealers are few, and as they are but beginning the honourable task of introducing the products of their native countries into England, and have to face an active set of competitors already in the field, it is right that they should be made known to the medical profession, through whom alone a salutary reform in the drinking customs of the country can be effected, and who alone can break through the prevalent superstition in favour of fortified wines and the fear of " acidity."

There are other Hungarian red wines, as the *Poganyvar*, of which samples were sent me in 1866 by Messrs. Homberg and Haas, including *Poganyvar auslese* at 30*s*.—a very agreeable, dry, well-flavoured wine; a Pogauyvar Schieler (*i.e.* a *squinting* wine, neither red nor white—clairette), a pleasant, strong, sweetish wine; and a *white Cabinet Poganyvar*, something like Mersault.

The Hungarian red wines seem as if they were the property of the enfeebled and phthisical part of our population. When a schoolgirl is brought to me with cough and chilblains I tell her to go back with a dozen of Ofner or Carlowitz, and a dozen of little tins of Gillon's essence of beef, and the prescription seldom fails. We shall have something more to say about wine for the consumptive, but, thank goodness, they suit the healthy as well. There are the superior growths to which Max Greger has affixed a gold label, and which I have known to win the favour of persons to whom light wines were an abomination, and to whom Hungarians were *vins sauvages*. A gentleman intimately acquainted with French and German wines vowed that the gold label Ofner was the finest Côte Rotie or Hermitage, and so it is in character.

Transylvanian Wine.—Transylvania is a country which, when means of communication are improved, will be able to send us excellent wine. Of these there is one, the *Mediasch*, which is supplied by Max Greger, and which seems deserving of attention. It is deeper-coloured, more acidulous, and of old Rhenish quality. A short time ago, a number of *savants* did me the honour to meet at my house to form a jury for the purpose of tasting and comparing all the artificially preserved soups in the market. I selected the *Mediasch* to refresh them under their labours, and they all approved it highly.

CHAPTER XV.

Austrian Vöslauer wines of M. Schlumberger.

THE Austrian wines furnish an answer to a question which is sometimes put to me. It is all very well, it is said, to bring these cheap wines into notice, but the moment a public demand arises for them they will cease to be cheap, for the demand will be greater than the supply, and then prices will rise. Or else adulterations and dilutions will be perpetrated which will disgust the consumer, and so the public will be worse off than ever. To which it is replied, that even supposing the vineyards of France, and Greece, and Hungary were to fail—stimulated as their proprietors would be by the diffusion of a greater taste and knowledge amongst the English—the Austrian vineyards at Vöslau would supply the deficiency. They belong to Mr. R. Schlumberger, who was one of the jurors of the International Exhibition, 1862, in London, and who has devoted his life to the introduction of the best vines, the best vine culture and wine-making into his vineyards. Large quantities of these wines are exported to Italy, the Danubian Principalities, Russia, &c., and it is my belief that they will meet with a steady sale in England so soon as they are sufficiently known. I am told that large quantities were taken in the Austrian frigate *Novara* in her cruise round the world, and that, after two and a half years, in a great variety of climates, that part of it which was brought home was found greatly improved in flavour.

Of these wines, some are still, some sparkling. Of the latter I must remark that the samples of the Sparkling Vöslauer which I have tasted, and which range from 46s. to 64s. per dozen, will hold their own against any of the liquids called "champagne" of equal price; and that a man who does not want to give an extravagant price for "champagne" will be well suited by Sparkling Vöslauer.

The Still Vöslauer wines are red and white; there is no complexity about them, and there are only three or four sorts of each.

The *Red Vöslauer*, the lowest quality, costs 15*l*. per hogshead in bond, or about 14s. 6d. per dozen, duty paid, but exclusive of bottles and bottling charges. It is 24s. per dozen retail; but I need not repeat that the man who buys in *quantity* may save 25 per cent. It is a good stout, full-bodied, serviceable, and, I believe, economical wine, as its stoutness renders it more satisfying than most Bordeaux of equal price. There is no complaint of thinness, sourness, coldness, or poverty; it is a good sound wine, with just roughness enough to be clean.

The *Vöslauer Goldeck*, at 25*l*. in bond, or 30s. per dozen retail, is a smoother, finer wine; whilst the *Goldeck Cabinet*, 36s. retail, is a much smoother, softer, more finished wine, which would be pronounced a "Burgundy," and would suit any roast meat at dinner, or might be sipped as an "after-dinner" wine.

What I have said of the red applies *mutatis mutandis* to the white. The *White Vöslauer* at 30s. is a good clean amber wine; very sound, not likely to offend John Bull by its acidity, and fit to appear at any dinner with fish and *entrées*, or at any evening party for young people *vice* firebrand sherry, or at the family dinner in hot weather of economical persons, who think 'it bad economy to deny themselves the means of healthy nutrition. The white *Vöslauer Goldeck* at 36*s.* is a better wine, and the *Steinberg Cabinet* at 42*s.* fuller flavoured, resembling a white Burgundy. The lower qualities possess *grapiness* (without too much perfume, without the muskiness of some light wines, which though agreeable in its place is not liked by every one at all times) with some vinosity; and some samples I have tasted of the higher kinds have a true Burgundiacal aroma. These wines are now sold by all the dealers in Hungarian and German wines.

I had the privilege of visiting Mr. Schlumberger in 1865, of seeing his vineyards at Vöslau and Goldeck, and witnessing the operations of the vintage. The richness of the over-ripe white grapes destined to produce the cabinet wine; the amplitude of the cellars excavated in the bowels of a hill; the vicinity of sulphur springs and volcanic débris, and the immense care, activity, and conscientiousness employed, bespeak a great future for these wines. We may say of them, in diseases of exhaustion and debility, what we have said of the Hungarian.

CHAPTER XVI.

A digression on mead or metheglin, with a few words on cider—The *Sicera*, or *Strong Drink* of the Bible—Decay of housewifery—Cases in which cider should be prescribed.

CIDER deserves a very few words in order to define the place it should occupy in the diet roll of the practical physician. Thinner, hungrier, and sourer as it seems to those who are accustomed to beer, these very qualities ought long ago to have ensured its use amongst a certain class of town populations—those, namely, who cannot afford wine, and who are becoming too heavy and corpulent. The acid and saline constituents purify, whilst the alcohol and aroma support and comfort. Many is the time that I have coaxed a patient into eating a dinner by proposing cider, although, through the unenlightened bigotry and spirit-loving propensities of many persons, they would almost as soon taste arsenic as cider. The dry cider—that which is neither too sweet nor too sour—should be selected.

The odd thing is, how all *à priori* conceptions of the effects of cider, as acid, &c. &c., are discomfited by empirical fact. I know two gentlemen intimately, of delicate digestion, tendency to headache, lithic deposit, and other indications of immature gout; one is a well-known F.R.S.: from them I learned the great digestibility of cider in such cases—even the very cases in which we should have least expected it.

In the next place, let me say a few words on that

ancient liquor called *Meade, Meth,* or *Metheglin*. I do not want my readers to drink it, but some account of it may be a contribution to that part of anthropology which consists in the history of fermented liquors.

An American friend once asked me if the English knew a drink called *cider;* thinking it was peculiar to America! (He sent me, by the way, a cask of superb cider from New York, which, after a long voyage, turned out particularly well.) Just so do I notice in Mulder's "Chemistry of Wine," that "honey wine or mead is prepared in Poland, Galicia, and some other parts (!!) from honey-water and ferment." Gracious heavens! is it come to this, that the drink of our Anglo-Saxon fathers, and of their British predecessors, which warmed them in fight and feast, and which they hoped to drink for ever in Hades out of their enemies' skulls—the true wine of the English yeoman, shall be talked of as if it were peculiar to Poland, Galicia, and "some other parts"!!

Mead is one of the oldest drinks in the world, as we before said. It forms one variety of the liquids classed together as *sicera* in the Vulgate, σίκερα in the LXX. and Greek New Testament, and under the name "strong drink" in the English Version. The Nazarites were forbidden to drink wine and "strong drink." Wine stands out by itself as the noblest of fermented liquors, as the highest gift of the kind to man, and as the type or symbol of the Divinest Influences that can be veiled under the Sacramental Elements. The "strong drink" or *sicera,* whence our word *sycer,* or *cider,* included every fermented liquor except grape-juice; such as palm wine, beer, cider, fruit wine, and mead.

Good mead is a liquid of very variable sweetness, according to the quantity of unfermented honey which may remain in it; if nicely made, it is nearly dry— *i.e.*, not sweet. By age it acquires a remarkably luscious perfume, like that of Tokay. I have examined many specimens, *ex. gr.*:—

1. Mead sent me 20 years ago by a medical friend in Hampshire. Most likely from having been boiled in an iron pot, it is so strongly impregnated with that metal that it has quite a chalybeate taste, and is undrinkable except to taste as a curiosity. Nevertheless, bottle after bottle has gone, as I have given it to some policeman or other person of West Saxon descent, who forgives the iron for the sake of the liquor. Carelessly corked, standing upright in my cellar for years, it is nearly dry, quite free from acidity, sound as possible, and has alcoholic strength 20·5.

2. A specimen about five years old, vilely made, full of unfermented honey, also standing upright in a carelessly-corked bottle; very sweet; sp. gr. 1080; alcoholic strength 18.

3. A specimen from a medical friend in Hampshire, made last year; sp. gr. 1020; alcoholic strength about 20; bright, clean, well fermented; strong tasting.

4. From a cottage on Poole Heath, made 1864; bad condition, actively fermenting, acid, sweet and heady; sp. gr. 1050; alcoholic strength 23.

5. From a cottage in Holt Forest, Dorsetshire, of 1864; clear and pleasant, not quite well fermented; sp. gr. 1027; alcoholic strength 24.

6. From a cottage on a heath near Cranborne, Dorsetshire; very clear, well fermented, and pleasant; too sweet for my palate, yet perfect as a specimen of

a sweet fermented liquor, and very fragrant; sp. gr. 1050; alcoholic strength 22.

7. A magnificent specimen from an eminent tradesman at Christchurch, Hants, made in 1814, and consequently more than fifty years old; sp. gr. 1080; alcoholic strength 16. Marvellously soft, full flavoured, and fragrant; a little drop perfumes a glass so that it is difficult to wash off.

Why, it may be asked, do I occupy my reader's time by descanting on these barbarous liquors? Because they tell us two things.

In the first place, they set aside the notion that any large quantity of alcohol is necessary for the maintenance, or preservation, or development of a fermented liquor. When we think of these specimens of mead that have been literally lying about, without any care, for periods varying from one to fifty years, and yet in perfect preservation; and when I add to them two samples of *elder wine* which I received from the same gentleman that gave me the fifty-year-old mead, one made in the year 1815 and one in 1818, each of low alcoholic strength, and yet perfectly preserved and marvellously nice, considering what they are—not to speak of cider from America tossing about for weeks in all weathers—when I add to these a specimen of Oxford ale that has been in my cellar for years upright in bottle, alcoholic strength 19—we may well demand from the wine-growers of Portugal, Spain, Sicily, and the Cape, that if they are to continue to supply the English market, they shall do upon scientific principles what the poor West Saxon peasants do by rule of thumb. Grape juice is but honey, and ferment, and water in a different shape; and what can be done with the one ought to be possible with the other. We ought to have firm and stable wines of the countries above named, without the addition of spirit.

Secondly, it is worth while for the medical philosopher to glance at the habits of a people, and at their luxuries, as evidences of their moral and social position. The cottager who can brew a small stock of mead, and keep it for feast days, or friendly gatherings, cannot be very low in the scale of humanity. He has evidently a little surplus, a little forethought, and some notion of those snatches of rest and enjoyment which distinguish the labouring man from the slave or beast. But I suspect that the custom of making mead is, like other branches of housewifery, dying out amongst the West Saxon peasantry. When I was a boy, brought up in a part of the ancient Wessex, a drop of mead was offered on calling at a better class cottage. Now, in 1865, it was with the utmost difficulty that the messengers who were good enough to undertake the task, and who trudged long distances over a country less visited than usual by change, could collect a few driblets of the liquor. Bees are more scarce; cottagers, if they keep them, sell their honey, and buy beer; but in all these matters *housewifery*, or the cost of keeping house comfortably, is dying out. Home-made and homespun are displaced by manufactures (or machino-factures) and shoddy. Formerly, *baker's* bread and *brewer's* beer were despised as unworthy to be set by the side of home-made; *butcher's* meat, too, was distinguished but as a thing of superior class from common meat—*i.e.*, pig meat; but the progress of events makes our whole population less housewifely, and more dependent on the shop. On this point Cobbett's " Cottage Economy" deserves to be attentively studied. Young cottage girls had better brew or bake than do crochet. Cobbett says, with more than his usual elegance, "Give me, for a beautiful sight, a neat and smart woman, heating her oven and setting in her bread! And if the bustle does make the sign of labour glisten on her brow, where is the man that would not kiss that off rather than lick the plaster from the cheeks of a duchess?" We, as medical men, may ask whether the woman who is accustomed to bake and brew, and who has a bottle or two of this fragrant honey-wine to set before a guest, is not more likely to be self-dependent, able to nurse the sick, rear a family, and pay a humble doctor's bill, than the woman who gets her cordials from the publican and her food from the shop, and who, when ill, goes straightway as a pauper to the parish or dispensary?

CHAPTER XVII.

Wine from Australia—Mr. Patrick Auld's Auldana—Mr. Wyndham's Bukkulla—Dr. Kelly's Tintara—Sound wine lore from the Antipodes—What missionaries should know about wine.

HEN Sir Morgan O'Doherty published his maxims fifty years ago, people were told that they ought to drink Cape wine because it was the only wine grown in His Majesty's dominions. Happily, there is now an abundance of better sorts; and he that desires to accomplish the patriotic feat of fostering a new enterprise set up by Englishmen of his own flesh and blood, and at the same time of supplying himself with some of the most excellent wine, should provide himself with some of the Australian.

Two hundred years ago, there was on the map a great blank, labelled *Terra Australis Incognita*. During the last ten years the great Australian colonies which now are peopling this void have raised their production from 444,917 gallons to 1,933,403 in 1870, and if they do but keep clear of the *oïdium* and other scourges, will probably increase at the rate of 1,000,000 gallons per annum. Thus they will supply their own great and growing communities, and export largely, as we hope, to the mother country. Nay, more, there is India. We hope the day is not far distant when *Auldana* and *Bukkulla* will be handed round at Government House at Calcutta and Madras, instead of the hock and claret now sent round from Europe.

My allotted space is running short, so I must

not indulge in any reflections on the past, present, and future of the wine-growths of that huge Continent; but will describe shortly the wines that are to be had here in London.

The first I must mention as grown upon soil inhabited by Englishmen, is the product of the Auldana vineyards, so called after their first planter, Mr. Patrick Auld, whose name will be handed down to posterity, like those of Dionysus and Ceres, inseparably blended with that of his excellent wine. The Auldana vineyards are situated near the River Torrens, above Adelaide, South Australia. They were planted by Mr. Auld at various times between 1837 and 1847, and have had his unremitting attention and endeavours since to make the produce as perfect as the purest juice of the finest grapes can be. The vines came from the stock introduced into the Australian continent by Sir Charles M'Arthur, and are selected and mixed in due proportions, some to give astringency, some flavour, softness, or strength. The soil is deep —the vine-roots penetrate for thirty feet through calcareous *débris*, with fragments of quartz and slate. It is reckoned that one acre of ground on the hills may produce 400 or 450 gallons of good wine; but if quantity be desired more than quality, five times that amount might be got on the plains contiguous to the river. The vines, which are propagated by cuttings, begin to bear at four years, and are in their prime after seven years.

The vintage is in March; the grapes are separated from the stalks by a machine, and are crushed by rollers covered with felt, so that the stones may not be cracked and so give their astringent bitterness to the wine. Every part of the process is conducted with the greatest cleanliness and delicacy—there is neither handling nor treading. Great attention and skill are required in the winemaker about the time of the vintage, as the grapes rapidly increase in sugar under a hot sun. Moreover, great care is required to insure uniformity of result, because in the same vat the upper layers of liquid may have their saccharine density reduced to 1°, whilst the under layers have 10°. The first fermentation lasts for six or seven days; the wine is racked in July, and twice a year afterwards. The wines mature rapidly under the great heat to which they are exposed, for they are kept in shingle-roofed houses, elevated about three feet above the ground, where the temperature is, in summer, from 120° to 160° in the sun, and often 90° inside. All the processes of winemaking have been made out and adapted to the local circumstances by Mr. Auld, who has been unassisted by foreign manipulation.

The wines, which have been imported into England, are to be had of Mr. Auld, Mill Street, Hanover Square, London, and of Messrs. Apps and Leigh Smith, Fenchurch Street. They are what the gipsy girl in "Peregrine Pickle" told Tom Pipes was "meat for the master"—that is, they do not aspire to be the *vin ordinaire* of daily humble life, but to be the restoratives of the invalid and the luxuries of people of condition. The white, which varies from 36*s.* to 50*s.*, is a charming wine, pale in colour, delicate in taste, fresh and fragrant as innocent girlhood, having no acidity, but a charming appetizing aroma of the fresh grape.

The red wines are lower-priced by one-third, and have a character of their own. A Bordeaux wine merchant called on me the other day, and, in order to improve the occasion, I had up a bottle of Auldana, and asked him what it was. He said that I must pardon him for his want of familiarity with Burgundy, but he supposed it to be *Pommard*, and greatly he stared when I told him it was Australian, for no Frenchman can conceive of any wine that is not French. But it is not like Burgundy either. From Bordeaux it differs in absence of astringency, and scarcely perceptible acidity, and in greater smoothness and rotundity; its flavour is *sui generis*, and taste very soft; and it would, in my judgment, suit any one who wanted a full-bodied nutritious wine without headiness.

So far I have written of bottled wine which reached me in 1869, and whose diminished ranks have improved in flavour to this day. Regarding later importations I can say, some of the white muscatel Auldana is truly superb, and in the fulness of its flavour reminds one of the old dry Ruszter. The red or ruby Auldana is a very fine wine; well matured, generous, not like claret, of which you should drink a bottle, but telling in a single glass; albeit a wise man will empty a bottle, if he has a friend to help him. We may say that all these Australian wines are at prices much below their merits.

Next in order let me speak of the wines of Mr. Wyndham, of Dalwood, Branxton, New South Wales, a man of the most ancient Wiltshire blood, who now is President of the Hunter River Vineyard Association. Of these wines, a Frenchman who visited me a short time ago pronounced the white Bukkulla simply perfect—a wine of exquisite finish; uniting the qualities of Hock and Meursault; the red is a good stout, full-bodied, nutritious wine, recalling some of the best wines grown upon the Rhine, and promising abundance of flavour. These wines are sold at a most moderate price, and both (bottled in Australia) are to be had of Messrs. David Cohen, 19, St. Helen's Place, E.C.

The *Tintara*; a third wine stock is the result of the labour (I may say the labour of love) of a member of the medical profession, Dr. A. C. Kelly, of Maclaren Vale, Adelaide, South Australia, and may be procured through Messrs. P. B. Burgoyne, of Old Broad Street. I have before me Dr. Kelly's book entitled the "Vine in Australia," published at Melbourne in 1861, a thorough treatise on vine culture, on the choice of vines and of soils, the management of vineyards, and the fermentation of wine, which may be taken as a guarantee that the writer thoroughly understands, and is determined to accomplish, the production of a good pure wine. The Tintara vineyard, which is on a gravelly and ferruginous soil, on hill slopes 600 feet above the sea, was planted in 1863, so that the *primitiæ*, or firstfruits (wine of 1867), of which I have received a sample, good as they are, will no doubt be surpassed in the future. The yield is, like that of the best Bordeaux, small, only 250 gallons per acre, so that quality is not sacrificed to quantity. The wine before me is quite mature, with something of tawny tinge; pre-eminently full-bodied and nutritious, with a port wine flavour. Like the Ofner and Carlowitz, these Australian wines are well adapted for convalescents. They are robust and satisfying.

I have since been favoured with some samples by Mr. P. B. Burgoyne, of *Verdeilho*, a light-coloured, fragrant sweet wine, from Gilbert's vineyard, at Pewsey Vale, near Adelaide. This wine is not to be had here at present, but I may notice it amongst the potentialities of Australian produce. Any physician who has a patient that requires eight meals a day may be grateful for the opportunity of using such a saccharine stimulant. But there are other Pewsey Vale wines to be had, both red and white. The "Pewsey Vale White, vintage 1864," is an example of the power of these wines, both in alcoholic strength and vinous flavour. The strength of this wine is equal to 25 per cent. of proof spirit, without evidence or suspicion of fortification; the vinous flavour intense, and the taste has qualities which approximate to those of a fine dry sherry—in saying which we pay a high compliment to dry sherry, for much of the liquid sold under that name tastes more of Glauber's salts than of wine. The "Pewsey Vale Red," also 1864, is a fine mature wine, grapy and potent, alcoholic strength 24°, fit to rank with Hermitage. The Tintara, I am glad to learn, is used in at least one hospital instead of the *banal* "Tarragona," or "Hospital port."

Any one who has the welfare of the human race at heart would read with delight a copy of the *Maitland Mercury* of May 11th, 1872, now before me, containing an account of a meeting of the Hunter River Association of Wine Growers. The accounts of the extraordinary fertility of the Dalwood vintage—30,000 gallons from thirty-six acres, the result of high culture, constant attention, and the trellis system—are most cheering. The remarks of Mr. Wyndham, the chairman, on the infamy of mixing crude spirit with immature wine, will, I hope, not be thrown away. He remarks that a few years ago, people imagined that fine wine would not keep without spirit, and that " to be worth drinking at all, it must be similar to port and sherry, and other spirit-loaded liquors to which the British palate had become accustomed during the last century. Since that time we have found all this to be a fallacy, and for years past our most strenuous efforts have been directed to the production of pure unbrandied wines; so much so that several of our most successful leading growers, myself included, have not availed themselves of the liberty to distil, nor taken out the license. Brandied wines are conducive to intemperance; and their use in the British nation and other English-speaking communities, has brought about the present odd phenomenon in the world of a large number of people banding together under the name of temperance societies, &c., &c., to prevent the use of wholesome food. All this is the natural consequence of the use of those abominable brandied wines. They are unnatural, and minister to the passions that degrade man. Pure unbrandied natural wines are conducive to temperance; and the fact that teetotal societies are confined to those parts of the world where brandied mixtures are in use, and unknown or almost unknown in all the great wine countries of Europe, where pure wine is the daily food of the men, women, and even the children, and drunkenness there being an unusual occurrence,—is evidence that *it* is the true and only specific for the present great evil of intemperance.

"A friend lately at Adelaide, who understands and appreciates pure light wine, and who is fully alive to all the benefits derived by a community where such beverages are in general use, as in European wine countries, was lamenting to me, the other day, over the evils produced by the use of young wines, loaded with abominable, crude, half-manufactured spirits, in South-Australia."

Mr. Wyndham's observations deserve to be reprinted and circulated by the Society for the Propagation of the Gospel and the Society for Promoting Christian Knowledge, to every clergyman and missionary at home and abroad.

CHAPTER XVIII.

Sparkling wines—Champagne, the great brands—Cheap sparkling wine from Saumur, Vouvray, Nêuchatel—The Styrian—Sparkling Hock, Moselle, and Tokay.

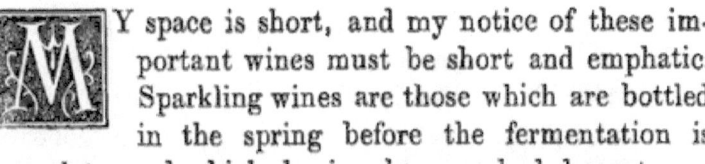Y space is short, and my notice of these important wines must be short and emphatic. Sparkling wines are those which are bottled in the spring before the fermentation is complete, and which, having been corked down to retain the carbonic acid, are suffered to clear themselves by further fermentation and deposit; then they are uncorked, and by ingenious manipulation disgorged from their sediment, and lastly are dosed with more or less of a *liqueur*, composed of champagne essence, or of brandy and sugar-candy; then corked and wired for use.

The finest wine of this class comes from the province of Champagne in France, whence most sparkling wines are graced with the name "Champagne." The test of excellence in champagne, or other sparkling wine, is that it should taste of *wine*; a clean, light, fragrant taste of the grape. Bad, dry champagne tastes of bad brandy, and bad, sweet champagne, of sugar-candy as well. Any wine in which these ingredients are prominent, will probably give rise to a stunning headache. So far as effervescence is concerned, the perfection of the wine is to have the carbonic acid so intimately dissolved that it escapes creamily, so that when the cork is drawn it does not blow half of the wine out of the bottle.

The uses of champagne in medicine are manifold. When, on an emergency, we want a true stimulant to mind and body, rapid, volatile, transitory, and harmless, then we fly to champagne. Amongst the maladies which are benefited by good champagne is the true *neuralgia:* intermitting fits of excruciating pain running along certain nerves, without inflammation of the affected part—often a consequence of malaria, or of some other low and exhausting causes. But there is another neuralgia, which is really a true rheumatic inflammation of some nerve, especially the sciatic, and attended with all the gastric and assimilative disturbance characteristic of rheumatism, and I can conceive of nothing more mischievous than the administration of bad champagne in such a condition. Yet I have known it done.

To enumerate the cases in which champagne is of service, would be to give a whole nosology. Who does not know the misery, the helplessness of that abominable ailment, influenza, whether a severe cold, or the genuine epidemic? Let the faculty dispute about the best remedy if they please; but a sensible man with a bottle of champagne will beat them all. Moreover, whenever there is pain, with exhaustion and lowness, then Dr. Champagne should be had up. There is something excitant in the wine; doubly so in the sparkling wine, which the moment it touches the lips sends an electric telegram of comfort to every remote nerve. Nothing comforts and rests the stomach better, or is a greater antidote to nausea.

I cannot descant on the great " brands" of champagne, which any wise man who has money to spare will prefer, unless he can rely on his own judgment or his wine merchant's. Let me rather give a word of comfort to people who want a sparkling wine, and yet cannot afford to pay eight shillings a bottle for Moet's, or Rœderer's, or Bollinger's. There is plenty of good sound wholesome wine to be had, and if not the best, yet quite good enough.

I have had from Messrs. Ackerman Lawrence, of Saumur, some capital sparkling wine, at less than three shillings per bottle. Tests :—It kept good three years in my cellar; and when in the winter of 1870 all my family had influenza — coughing, sneezing, crying, wheezing, &c.—a good dose of this wine was "exhibited" all round with the best effect.

There is a wine called *Vouvray*, imported by Mr. Arthur Browning, of Lewes; a good honest light wine, of which he was obliging enough to send me samples in three conditions :—First, as a finished still wine; secondly, as a sparkling wine unaltered; thirdly, as a finished *vin mousseux* with liqueur. The wine was thoroughly wholesome, and very agreeable.

Excellent sparkling wine comes from Neufchatel; there is a Styrian champagne of the Brothers Kleinoscheg, of which Bergmann and Co., 32, Great Tower Street, are consignees, and very good it is; sparkling Burgundy, Vöslau, Tokay, Hock, and Moselle; sparkling Vino d'Asti; sparkling Rheingau.

In addition to its uses in sickness, let me say that a good inexpensive sparkling wine is the best and most economical for the entertainment of large parties, balls, &c., and that at a dinner-party it would be liked better by many people than wines worth four times the price, although the philosophers ought to have the choice of the better wine.

CHAPTER XIX.

Fortified wine—Rogomme, Port, Sherry, Madeira, Marsala, Cape—General characters—High art and low art—Spirit added to Port; its effects—Characters of good Port—Tarragona, Roussillon—Characters of good Sherry—Dry and sweet—Sickly and sad Sherry—Hypocrisy and tyranny of Fashion—Madeira—Marsala, Malaga, &c.

THERE was an amusing book of Travels in Italy, published in the last century (I forget by whom,) in which the writer describes his efforts to get his servant, a good, simple, English countryman, to appreciate the works of art with which that favoured country abounds. They were staying in some Italian city (I forget which), whose galleries are renowned over the civilized world. The travelling connoisseur in vain tried to get his servant to accompany him through saloons where the majesty of Michael Angelo, the grace of Raphael, the quaint simplicity of Peter Perugino, the sublimity of Murillo, the tenderness of Guercino, the unctuosity of Carlo Dolce, and the pure, unconscious nudity of antique statuary shone supreme. No; the honest fellow could never be got farther than the vestibule, where his satisfied gaze was riveted on a flaming picture of Joseph and Potiphar's wife.

Oh, ye country clergymen! Oh, ye British fathers of families! Oh, maiden ladies of immaculate propriety! pardon me I pray you, if I tell you that your behaviour with regard to wine is like that of the yokel in this parable. There is the sumptuous Burgundy, the graceful Bordeaux, the nutritious Hungarian, the sturdy Australian, the robust and wit-compelling Rhenish—for all these ye care not; you never try how excellent are the wine flavours; you are satisfied at the outset with the coarse and spirituous; so as a wine be strong and fiery you want no more. "Wine," you say, "should do us good"—that is, burn your stomachs.

The alcohol which you prefer to wine is really a mere drug, a cheap and worthless product, whose money value in England is from 1s. 11d. to 2s. per gallon. Its cost is due to the 10s. which it pays to the revenue.

But the predilection for alcohol exercises the most debasing influence on wine culture; it enables wine-forgers to concoct any variety of fictitious liquids which the public will buy and drink if strong enough, and for these it will pay exorbitantly.

For instance, some casks of red Australian wine came to this country damaged by leakage. What was done with it? It was taken to the wine forgers at Hamburgh, who sweetened, drugged, and fortified it, and then sent it back to Australia as *port!* which the ignorant colonists would pay for at three times the rate of their own pure, unadulterated wine.

M. Jules Guyot describes a liquor called *Rogomme*, which he met with in his journey amongst the wine districts of France, made of the juice of the most luscious grapes and of brandy. This is the type of fortified wine;—spirit and unfermented, or half-fermented must—and every grade is to be found, whether as to the degree of fermentation, or to the quantity of added alcohol.

The following statement may be relied on as an

account of the composition of port wine of the first quality. It was given to me by one who has a better right to know than most men:—

"COMPOSITION OF PORT WINE OF FIRST QUALITY.

"To the pipe of half-fermented must is added, to check fermentation—

	25 galls. of brandy	= 37·5 proof galls.
Say	5 ,, elderberry juice to colour.	
	6 ,, more of brandy	= 9· ,,
	2 ,, ,, after racking	= 3· ,,
	1 ,, ,, on shipment	= 1·5 ,,
	39 liquid galls.	51· ,,
	76 of wine.	

115 galls. of port wine.

"Taking the probable strength when half-fermented at 14° (the highest natural strength known being 28°), the strength would thus be about 42° or a little above it."

In considering this statement, let me ask my readers to ponder on the quantity of alcohol; the quality thereof; the price; the effects of alcohol upon wine, as wine; and the result as a beverage for civilized beings.

The *quantity* is such that a glass of port wine equals in strength more than two-fifths of a glass of brandy.

As to *quality*, the added spirit may be English or foreign; distilled from grain, sugar, beetroot, or potatoes, and contaminated with those *feints* which have the most deleterious effect on the nervous system. "Ridley's Monthly Circular," for March, 1865, says of the exports of British spirits during 1864, "Our best customers are the Portuguese vine growers, who have taken upwards of 1,500,000 gallons to fortify their unfermented juice."

In 1864 we took from the Portuguese 3,344,871 galls. port wine.

,, They took from England . 1,630,304 galls. of spirit.

He must be dull indeed who does not see that in paying high prices for port wine we are really buying back dearly the British spirits that were engendered on the banks of the Thames.

Moreover, if we consider the composition of port, which costs, say 5$s.$, one-third of the bottle will be spirits, worth $1\frac{1}{2}d.$, whilst the remaining two-thirds of *wine* will cost 4$s.$ $0\frac{1}{2}d.$, allowing 5$d.$ for bottle and cork, and 5$d.$ for duty; total 5$s.$

The *effect* of alcohol upon the wine is, first, to check fermentation and preserve a certain lusciousness; but then the wine, imperfectly fermented, tends to ferment again, unless held in check by further doses of alcohol. Secondly, the addition of spirit really robs us of so much wine—we lose the virtues of the grape juice. Thirdly, the effect of spirit upon wine is to kill it as wine; to precipitate colouring and extractive; to make it prematurely old and tawny, without the precious perfume which really old wine has. It makes the wine turbid; precipitates albuminous matters, which ought to have been got rid of by fermentation; and when they have settled, and the wine is racked off, it is much better able to bear carriage. We can now understand how *fruity* port can be rapidly transformed into the *prime port, old in the wood, tawny, and very dry,* which is sometimes offered to the ignorant at low prices. A good dose of spirits and a little wine will make a tawny liquid, thin to the palate, but fiery to the throat, with scarce a smell of wine in it, and with little, if any, crust on the bottle or cork, which there must have been if the wine had grown old, and deposited its extractive in bottle. The phrase *old in the wood* applied to cheap port really means this:—Here is a liquid with little extractive matter. A man of common sense, who knows that port wine ought to have a good deal of extractive, will naturally look to the cork and the bottle to see if it has been deposited there as a crust by lapse of time; and if so, he will naturally expect good flavour, softness, and no prominent alcoholic taste. But this cheap dry port has no crust and no flavour, and a *very* prominent alcoholic taste. The wine merchant vows that it dropped its extractive through age in wood. *Credat Judæus!*

Then as to the effects of port wine upon civilized man; if there be one fact better proved than another, it is that it cannot be drunk habitually in small quantity, nor yet be resorted to as an occasional luxury in large quantity with safety, by large numbers of our population. Ask any four men of forty and over, and three will say, "I can't drink port; I am afraid of the gout." It is curious, on looking through the evidence given before the Committee of the House of Commons on the Wine Duties, in 1852, to see even then how the veteran wine merchants and œnologues who were examined, spoke of the taste for port wine as "vicious," and as "declining." Mr. H. Lancaster spoke of the consumption then as declining year by year, in proportion to the population. Mr. T. G. Shaw, in his "Wine, the Vine, and the Cellar," described the taste for port as unmistakeably going down in 1852, and so did Mr. Cyrus Redding, author of the well-known treatise on Wine.

Still there is no denying the fact, and it would be very ungrateful to deny it, that the old port wine was an admirable medicine when we wanted to pull a poor wretch up out of a fever or other state of intense debility. Good port wine of the right sort, which answers to the description given it by Forrester—" lively and clean on the palate, dry-flavoured, with an enticing bouquet; colour varying from pale rose to bright purple; perfectly transparent and mellowing with age, the rose becoming tawny, and the purple ruby, both of which colours are durable"*—such port wine, I say, though good as an alcoholic medicine, yet if of higher alcoholic strength than 30, was not fit for the ordinary drink of healthy persons.

It is true that the quantity of port wine imported has increased of late; but it goes to a different class from those who formerly used to drink it: instead of the House of Lords and Bench of Bishops, or her Majesty's Judges,—now farmers, mechanics, and *nouveaux riches*, who don't know better, are the great patrons of port. It is the upper classes who have substituted the more refined light wines.

The ill effects of fortified wines are twofold: first, the headaches and dyspepsia which arise from the crude added spirit, which are not caused by the natural spirit developed in well-fermented wine; and secondly, the *gout*, which is the offspring of those half fermented matters which the spirit keeps in check. Spirits have enough to answer for—nutmeg liver, delirium tremens,

---

* A Word or Two on Port Wine. Edinburgh: Menzies. 1844.

dropsy, Bright's disease, and the like. But much as spirits have on their heads, no one accuses them of causing the *gout*, but this is notoriously the offspring of strong, sweet, ill-fermented liquors—of which port is the chief—as ale, stout, and fortified wines. This is the opinion of Charcot, Ball, and the ablest Paris physicians, as well as of their English brethren.

What the physician desires to find in Port wine is, that the taste of spirit shall have softened down by age; that the luscious sweetness shall have subsided; and that there shall be the true wine flavour, which in the older and better vintages is unmistakeable and excellent. If old port wine possess true vinous fragrance it is probably wholesome; and it is the absence of this fragrance which damns its false imitators.

Cheap substitutes for Port.—Of these the best is Masdeu; the noisiest is *Tarragona*, or Spanish Red. Next come the fortified red wines of the South of France, of one of which M. Guyot says, with honest shame, that it is deplorable to see what ought to be the stout, sound, full-bodied wine of Roussillon contaminated, not with spirit of *wine*, but spirit of beetroot, and spirit of maize. "I have," he says, "seen under my own eyes, and under the inspection of the *douane*, 15 per cent. apiece of maize spirit added to more than 100 *muids* of wine on the quay at Port Vendres. *'Tis true, it was for exportation.*" But, he continues, how deplorable to see the ancient reputation of a whole province sacrificed by greed! Of the Tarragona class, it is enough to say that they are flat, more or less luscious, and destitute of wine flavour. This is the case with the South African. Other compounds there are made up, which have a treacly taste of some boiled syrup. The addition of boiling water detects these concoctions. There arises none of the vinous fragrance of true port. Good whisky punch would be honester and wholesomer.

Now for *sherry*, under which term are included, in popular language, all the white wines which come from Spain, and others like them. Monotony and base servile imitation are the curse of English life. People are expected to do, wear, eat, and drink things, not because they are pleasant or good, or useful, but because other people do it. Tourists in England must carry a chimney-pot hat to go to church in. Your own shirt-collar and beard, and your wife's bonnet and drapery, are settled for you by an irresponsible public opinion. The same with what you eat and drink. The fish, entrées, &c., must be accompanied with the inevitable sherry. All the fun, and the fragrance, the gratified sense of novelty, the curiosity as to the great political and social fortunes of our colonies, which would be excited by handing round a bottle of white Auldana; all the sympathy for our dear neighbours which would be excited by the taste of Meursault Blanc; all the respect for the Germans which would follow a sip of Hochheimer; all the hopes and fears felt for the Austrian empire, which would go round with the generous Vöslau, are smothered by the monotony of the *banal* sherry. When people are doing the serious act of dining, they should do it, and think about it, and talk about it; but to talk there must be novelty, not one dull perpetual round, and sherry gives rise to no ideas. England will never be merry again whilst it sticks to so sad a drink.

The physician who prescribes sherry, will look for a bright amber fluid, adhering slightly to the glass, smelling of wine, and tasting of wine. In fine old dry sherry, I have been struck with the perfect preservation of the smell and taste of grapes. The spirituous heat ought to be softened; and there should be abundance of the sui generis taste of old wine, making a soft, rich, aromatic, clean, fruitily fragrant whole. Taking this as a starting point, we may recognise the leading genera of sherry diverging in opposite directions. The one kind increases in colour from golden to dark brown, and in lusciousness; but it is perfectly clean, although in the dark brown sherry the taste of boiled must is unmistakeable. The other kind increases in dryness, and in ethereal fragrance, and lightness of body, passing through the grades of dry, or fine dry sherry. There is a wine of laudable fineness and purity called *Montilla*, and wine like it is called *Amontillado*. There is also a peculiar flavour called the "Amontillado," which probably arises from the presence of aldehyde, and which is singularly developed in some white Greek wines. There is also a wine called *Manzanilla*, introduced by Dr. Gorman, of a peculiar light fragrant and bitterish taste. *Vino di pasto* means a light breakfast wine.

Waiting upon these, like parasites or satellites or apes in human garments, are two classes of bad sherry, which the physician may call the sickly and the sad. By sickly sherry are meant those detestable liquids, hot, fiery, and yet sickly-sweet, that are advertised incessantly as the "Marylebone," the "People's Sherry," the "Lord Mayor's Own," &c. &c. By sad sherry, is understood a dry liquid, often hotter than any spirit of wine could make it, thin, with little body, but with a smell like that of nitric ether; so dry is it that there is no wine to be tasted in it. There is a thin sad stuff, which often borrows the name of *Manzanilla*, bitter, indigestible, and acescent in the stomach, said to owe its bitterness to chamomiles which grow near the vines, but in reality to spoiled wine pickled with spirit. There are some natural or unbrandied sherries of low price, which keep very badly, are very thin and flat, have no body and are not wholesome, and seem to have had their acidity neutralized by artificial means.

Of sherry in general, there is no doubt whatever that after it has undergone a certain amount of fermentation, it receives an addition of spirit—said to be six gallons per butt, and another four on being shipped. Flavour, softness, colour, richness, and piquancy are the product of the addition of portions of a mother wine—old brood wines kept in large butts which have undergone a special etherification, and communicate it to other wine. For richness and dark colour, a stuff called the "Doctor," composed of wine made from juice concentrated by boiling. For piquancy, an admixture either of the thin, dry, light, pure Montilla, or of Amontillado.—See Shaw, *op. cit.*

The best account of sherry is that given before the Committee of the House of Commons on the Import Duties on Wines in 1852, by Dr. Gorman, Physician to the late British Factory at Cadiz, long a resident in Spain. He says that no natural sherry comes to this country; it is all mixed and brandied. The quantity of proof spirit which good pure sherry contains by nature is 24 per cent., possibly 30. The less mature and less perfectly fermented the wine, the more brandy is there added to it to preserve it. Yet let it never be forgotten, Dr. Gorman added, "*It is not necessary to infuse brandy into any well-made sherry wine; if the fermentation is perfect, it produces alcohol sufficient to preserve the wine for a century in any country.*"*

---

* Minutes of Evidence, part 2, question 5776.

Now as to the uses of sherry. Good old sherry is a most valuable cordial and stomachic, and has marvellous uses for stimulating a feeble heart and making a refractory stomach do its work. And if patients or physicians desire to take or prescribe sherry, let me say with that great animal, Martin Luther, *Pecca fortiter*, do the thing handsomely; keep a reserve of fine old sherry for infirmity, and use it thankfully.

But for any young man whose stomach is sound, and who wants wine for its refreshment and flavour, to go to a cheap, sickly sherry is to be gouty before his prime, and to go to a thin sad unbodied stuff, is to sin without any fun; you drink the flat stuff, but I defy you to say you like it. If you desire a white wine, look up the vintages of France or Hungary—don't drench yourself with this miserable ghost of "Spanish white."

Persons troubled with acid dyspepsia and the allied diseases, gout, rheumatism, sick headaches, &c. &c., must drink either the best sherry or none. They should try a fine light Bordeaux; if they cannot digest wine, cold weak brandy and water.

Madeira does not rank with cheap wines; it sells at a high price, because generally kept till very old, and is valued the more if it have undergone a voyage or two to dissipate the heat of the added spirit and produce mellowness. It abounds in the truest vinous elements; it is a most potent stimulant to the nerves. Its ethereal elements are peculiar, and so good that it is a praise to any old wine to have Madeira flavour. Some acquires a flavour intense and singular, which is likened to *cockroaches*. Mr. D., who sent me specimens of this amongst other superb old wines, may be assured that his good seed has fallen on a grateful soil.

Of substitutes for sherry, the honestest is the Sicilian Marsala, a wine which I have tasted of exquisite purity and softness of flavour at the house of T. B. Unfortunately, most of it which reaches this country is fortified to the most atrocious extent. Yet if a man will drink strong wine, or if he desires to have a serviceable wine of the sort in his house, let him lay down Marsala instead of cheap sherry.

When old, although it preserves a coarse, earthy *taste*, it acquires *bottle flavour*, so much as to astonish persons who taste some forgotten bottle of it that may have been lying for years in the cellar.

Bucellas was a most useful wine, of high vinous character, ten years ago; now it is seldom worth drinking.

Malaga is a sweetish, flavourless, fortified wine, which has a great and utterly undeserved reputation in the South of Europe as a restorative and nutrient. Teneriffe and Lisbon are sweetish fortified wines. The wines of the Cape of Good Hope, or South Africa, seem to have dropped out of sight of late. They seem to have good material, but not well made use of.

I may wind up the subject of fortified wines with the following few remarks :—They are expensive when good, and suspicious when cheap. They may introduce elements foreign alike to good wine and to good brandy. When old and good, they often suit the aged and dyspeptic. With regard to hospital and charitable uses, suppose cheapness an object, the medical attendant will determine whether it is a stimulant or food which is wanted. If a stimulant, some clean spirit, mixed, sweetened, and flavoured; if food, a good young red wine, as Tintara, Como, Carlowitz; if the blood seems out of order from scurvy, some old thin red wine. Ladies need a good deal of education about wine; they should be taught the difference in nature, flavour, and effects; highly sensitive and gifted as women are with perceptivity of odours, they have never been taught to look for the juice of the grape and its admirable bouquet; they know only the effects of alcohol, and as it would be considered a scandal to drink many glasses of alcoholized wine, and as some houses have only the miserable little wine-glasses adapted for those wines, and as such a glassful of "claret" would be cold and flat, so they prefer something stronger. When Monsieur Assolant visited England, at the time of the International Exhibition, 1862, he caused wondrous offence by describing the English "Miss" as fond of brandy. The fact is, that our sherry would be called brandy by a Frenchman.

It is the hot sensation of spirits which is the delight of the lower orders; and this should be discountenanced as much as possible by those who know better.

Lastly, I must allude to the wine forgeries which are perpetrated at Hamburgh, producing, to quote from Ridley's Circular, Jan. 12, 1872, "an objectionable compound of neutralized acid wine, Elbe-water, potato-spirit, capillaire, and chemical flavouring matter." Ridley's Circular states that the importations are declining, and that no respectable house will now touch this pitch. *Elbe Sherry, Vatted Sherry, Hamburgh Sherry*, are the common designations; but any dealer may call it what he likes. In my former edition I treated largely of this; now I must confine myself to what is better.

CHAPTER XX.

Derogatory estimates of wine—Laputan Philosophy—How to treat a cask of wine—Order of wine at dinner—Wine for the consumptive—Wine duties—Gratiarum actio post vinum.

EROGATORY Estimates of Wine.—Wine is a unit, differing from all other things, and to be valued for its own excellence. Foolish and officious chemists sometimes foist their opinion on unsuspicious wine merchants, and pretend to find the explanation of the value of some wine in the predominance of some given chemical element. Thus it is distressing to read in the otherwise just and admirable eulogium of Dr. Le Gendre on the wines of the Médoc, that they excel because of the "action reparatrice de son tartrate de fer." In like manner an absurd report was spread about the Hungarian wines, that they contain phosphorus, in which case they would have the abominable odour of lucifer matches; or, again, as M. Kletzinski affirmed, that phosphate of lime and iron were the source of their virtues. Wine merchants are not responsible for this nonsense.

Phosphate of lime, or bone-earth, a compound of phosphoric acid and lime, and iron, are both necessary ingredients in the framework, the skeleton, the scaffolding of all things, animal and vegetable; and they abound in the coarser mechanical parts; in the bran and chaff that protect the grain more than in the delicate flour inside; in the bones more than in the nerves. Now a certain quality of bone and coarseness is needful in order that anything may stand upright: children may be too finely fed and bred, without enough of the coarse elements; then various kinds of ill-health arise, in which the admixture of the coarser elements of bran and oatmeal is beneficial. For common family use a coarsish wine is more sustaining; and the coarser the wine the more phosphate and iron does it contain. Compare the two following analyses made for me by Dr. Hofmann, late the able Director of the College of Chemistry:—

Max Greger's Carlovitz, selected, at 32s.

	Grammes.
I. Total solid matter (dried at 110° C.)	2·2720
Ash	·29995
Phosphoric Acid	·04162
Iron (met.)	·0027

Denman's Como, at 30s.

	Grammes.
II. Total solid matter (at 110° C.)	8·0216
Ash	·5201
Phosphoric acid	·0735
Iron (met.)	·0034

Here are two wines at about the same price; one a finer matured, lighter wine; the other newer, coarser, sweeter; and the latter shows half as much again of mineral ingredients—ash, phosphoric acid, and iron—as the former.

Laputan Philosophy of Wine.—When a tailor wants to fit a man with a coat, he takes a tape and measures him, and cuts his cloth accordingly. But this is too simple for the philosophic tailors of Laputa. They take the man's altitude with a sextant, his width with a geodesical goniometer, make a trigonometrical scheme, work out sines and tangents in logarithms, and send in a terrible misfit. Just so with wine. Do you want to know if it is good? taste it. Does it agree? drink a bottle at dinner, and then you will get knowledge truly empirical, positive, and not to be disputed. But the Laputan chemist amuses himself by calculating various ingredients to the fourth decimal, and then ventures to recommend the wine because of its proportion of alcohol, or acid, or the like. This is to substitute a roundabout way for a straight one, and to trust half a dozen hypotheses instead of one direct experiment. It is as sensible as to estimate virtue by weight or wisdom by measure. The chemist, as Mr. Griffin wisely said, cannot say whether wine be good or wholesome. A good wine may possess such and such elements in a given proportion; but it does not follow that any wine with the same elements in the same proportion must be good.

How to treat a cask of wine.—Immense is the rise in general culture, sweetness, and light which is indicated when any family or *côterie* instead of ordering their wine by driblets, have a cask direct from Bordeaux, Beaune, or Mainz, and bottle it at home. It is an incident almost comparable in interest to the birth of a baby, and fully as instructive. The cask will be looked upon by a wise parent, not as a hooped receptacle of drink, but as the "potentiality" of happiness, health, and good temper for his family. He knows that all substances exposed to air are wetted by it,—*i. e.* that the air sticks to and is imbibed by their surfaces, just as things are in water; so he will be careful to put the cask in some clean, airy, not close nor damp place, neither too hot nor too cold, and not in a draught. The cask must be put on something firm, with one end tilted. The bung, covered with a piece of tin, must be uppermost. Having thus put the cask into the position it is to stay in, the next thing is to fine the wine. For this purpose take out the bung; it is easily loosened by rapping the cask all round it with a hammer. Bore a hole with a gimlet in the head of the cask, and draw out a pint of wine. Close the hole by driving in a bit of pointed stick. Don't taste the wine, or you will think it horribly rough and nasty. Take the whites of half a dozen eggs (for a cask of forty-six gallons of red wine), and beat them up with a pint of wine thoroughly. Then pour the mixture into the bung-hole, and stir up the whole contents most thoroughly with a clean stick or yard measure; then fill up the cask with any good wine of the same sort if you have any; drive in the bung, and anyhow turn the cask on its axis, so that the bung may be under the wine. The Burgundians advise, that if you have no wine you should fill the cask with clean stones, so as to shut out all air. Let the wine rest a month; meanwhile collect bottles, have them well washed and drained, and see that no shot be left in them. Get corks to match, and let them be good ones. Don't have bottles with very large necks. When the wine is quite clear fill the bottles and cork them tightly; and lay them down so that the wine shall quite fill the neck and cover the cork.

It is quite a part of woman's mission to learn such matters as these; and the wine that you will put on your table, saying, "This is the *Thorins* '65, which we bottled ourselves;" or, "Take a glass of our own *Côtes de Fronsac*," will be relished the more because it will savour of its mistress's activity and skill in housewifery.

Order of wine at dinner.—This, according to Brillat Savarin, should be from the simpler to the more potent. For a small party of friends, who really want to *dine* seriously, they may (if they choose,) begin with a thimbleful of sherry or Madeira; next with fish and entrées, a good white wine, Graves, Pouilly, Niersteiner, Scharlachberg, or Œdenburg, St. Elie or White Auldana; with the roast, the best red wine in the house, Volnay, Côte Rotie or good Bordeaux, Assmannhauser, Red Vöslau, or Gold labelled Ofner, or Naussa; then some sparkling wine (this should be handed to the ladies throughout); at the end of the dinner some fine Rhine wine or Bordeaux. Then coffee, clear, hot, and black.

Does it hurt to mix wines? Not if each be good.

Wines for the Consumptive.—Oh, ye Ladies Bountiful, ask a schoolboy to translate the Latin Proverb, "vacuisque venis, nil nisi lene decet." If you give wine let it be the lightest unbrandied Ofner, Fleury, or Beaujolais, or Bordeaux; don't set the evening fever and cough going with a hot alcoholized wine such as Port.

Wine Duties.—Free trade is not only true policy, but being so, is a part of true religion. We must have taxes, and of these, large quantities are raised by putting 10s. a gallon on spirits of proof strength. *Natural* wine which is assumed to contain 26 per cent. of proof spirit or under, pays 1s. a gallon; artificial wine with from 26 to 42 per cent. of proof spirit pays 2s. 6d. per gallon duty. Some personages are agitating to make the duties on all wine equal; so that a person who would pay 2l. 10s. duty on the distilled spirit in five gallons of gin, may pay positively nothing for the same distilled spirit when existing as an added ingredient in 12½ dozen of port. This is a return to ancient protectionism, and is proposed as a bribe to the Spanish and Portuguese, in the hope that they will buy more largely of Manchester goods; in other words it is a

fined it? or was it because of the dark and evil times mild and good men were forced to shelter themselves in cells, and keep their liquor under lock and key?

However it may be, those cowled, good fellows had a good time of it. Safe and sure, tranquil and content, they fermented good wine, and brewed bad metaphysics; the wine they kept to themselves, but the metaphysics they let loose in the world to bother mankind, and deepen still more the darkness of the ages.

Soon after leaving Capri, I spent several weeks at the neighboring island of Ischia. At both places the vine-culture was substantially the same as at Sorrento, and at both places they cured oïdium unfailingly.

At Rome, a gentleman, whose father was a large owner of vineyards, gave me several kinds of wine to taste. All were decidedly pleasant, and the Aliatico delicious, but most of them having any age were more or less pricked. This gentleman, in giving some details of their modes of cultivation in the Roman territory, remarked that, for heavy work—trenching three feet deep, for instance—the most reliable laborers were from the province of Naples, describing them as strong, willing, and of excellent conduct.

From Rome I traveled northward into Tuscany,

where cultivation in all branches is thorough, systematic, and careful, and there I found no vines trained either on stake or trellis; all were clambering in tree-tops. Twenty-five feet was usually the distance between the trees on level ground, and fifteen feet on hills. Two or three vines were planted at the foot of each tree. This system is not confined to Italy alone; it is practiced in portions of France also. In the north of Italy it is common to prune the trees, so as to let in air and sunshine, while in parts of the south care is taken to keep them shaded. We often hear of vines grown upon trees in our own country, which, for some reason, escape disease, and from such facts an argument is drawn in favor of long and high training; but the immunity is probably due to the shelter from radiation which the foliage of the tree affords. M. Du Brieuil tells us vines trained upon trees in France suffer more than those on stakes. I learned the same thing to be true of trellis-grown vines in Burgundy. We know that in Italy neither trees nor trellis avail aught, and we shall find that in Southern France the lowest vines are least afflicted, and the highest suffer the most.

I left Italy by a wondrous road which skirts the Maritime Alps on one hand, and the Mediterranean on the other, and is called "Riviera" at one end, and

"Cornice" at the other, traveling in a carriage hired for the whole journey at Spezzia, where it begins.

It is a journey of six days, but so varied and so beautiful in all its ups and downs, ins and outs, that when it ends one is tempted to turn about and go back over it again. Now we descended to the very edge of the sea, and traveled for long reaches on its pebbly or sandy beach; now, mounting high, were whirled at a gallop along the verge of a precipice; now we rounded a rocky cape, on whose bleak sides no plant could stand, and now turned into a cove luxuriant with olive-trees and vines. Those who love the blue Mediterranean may thus, curving about her shores, embrace her, as it were, in a delightful week of prolonged leave-taking, and part from her at last more in love than ever. For nearly the whole distance the abrupt sides of the mountains were terraced with walls of stone, almost from foot to crown, and the soil thus secured planted in vineyards and olive-orchards. Much as we praise the Hollanders for building the dikes which keep back the sea from coming in upon their lands, the Italians deserve scarcely less credit for those dikes of stone which keep theirs from tumbling down into the water. As to the terrace-work of the Rhinelanders, it is as nothing in comparison.

CHAPTER XVI.

THE SOUTH OF FRANCE.

AT Nice I entered on the great vine-country of Southern France, where an enormous quantity of common, and a moderate quantity of superior wines are produced. In this region — namely, at Nice, Nismes, Montpellier, Cette, and other places — I remained about six weeks, with two subjects of inquiry in view—one, the vine disease, and the other, training *en souche*. What I learned on these points elsewhere I have mainly reserved for this place in my book, because in Southern France it is that the disease has been the most virulent and been most triumphantly subdued, and there it is that from time immemorial all the vines have been kept *en souche basse* (on low stocks).

The vine region in question extends from Nice in the east to Leucate in the west, and lies mostly between the 43d and 44th degrees of latitude, though extending as far down as below the 43d degree on the western wing, and, where the valley of the Rhone is included, going nearly as far northward as the 45th. It reaches from the Alps to the Pyrenees, and includes the entire French Mediterranean coast. It is sheltered from the winds of winter by a line of mountains that bound its whole northern border, and from whose bases the whole surface slopes gradually down to the shores of the sea. Composed of portions of ancient Languedoc and Provence, it includes the present departments of Drôme, Ardèche, Vaucluse, Basses-Alps, Var, Bouches-du-Rhone, Gard, Herault, Aude, and Pyrenées-Orientales. Of its entire tillable surface, fully one fourth is in vines—namely, a million and a half of acres, and the culture is continually extending.

The formation is generally limestone. The soils are various. On the poor slopes at the base of the mountains very superior wine is grown. Below them, at different stages of elevation, but mostly of level or slightly-inclined surface, are strong, but not over-rich soils, clayey, limy, and sandy in different proportions, capable of yielding large crops of strong, sound wine, which sells, when new, at from ten to twenty-five cents a gallon. Here and there on the level ground are found pebbly deposits whose product, like that of the poor hill-sides, is of a high order. Finally, the rich alluvial borders of the rivers have been known to produce, per acre, in a favorable season, as much as 4000 gallons of weak wine, containing only six per cent. of alcohol, formerly destined for the still, but of late years used to compound with other sorts in making cheap wines of commerce.

Observations taken at Montpellier show the climate of the region I am trying to describe to be marked by strong peculiarities. The mercury rises above 86° Fahrenheit, on an average, 34 times in a year. There are, in a year, 174 fair days (at Paris there are only 56). The mean number of rainy days in a year is 81. The yearly rain-fall averages 924 millimetres, about 36 inches. Rain often comes in torrents, as it does in America, but does not any where else in France. But little rain usually falls between the middle of June and the middle of October, an advantage which is somewhat compensated for by the heavy rains in the last half of October.

For the reason that the large average rain-fall of 36 inches is poured down in comparatively few rainy days, and the farther one that the prevailing winds are mostly violent and drying, the climate is a very dry one.

Although the mountains on the north are a shelter against cold winds coming from beyond them, they themselves give forth the frequent, sudden, violent, persistent, and biting cold "*mistrall.*"

Thus we find in the south of France intense summer heat, sudden and rude changes of temperature, high winds, heavy rains, and great dryness of amosphere, making up a climate very much resembling our own, and very little resembling that of any other part of France or of Germany.

When I first visited Languedoc, all its broad fields were in the plenitude of their autumnal array, the vines wearing their green, their purple, and their pearls displayed on outspreading and low-trailing branches, as if each one were a belle or a bay-tree. The next time I saw them, which was in March, and after winter-pruning, nothing met the eye but little, low brown stocks ten inches high, with all their branches cropped to two or three inches. Crinoline had given place to fact.

THE OÏDIUM.

Maybe many of my readers will think the pages given to this subject contain nothing important for them to know. Let them not be too sure of this. There may be regions in America, as there are in Europe, where the scourge has not yet come, and may never come, but such will be exceptional, and we can not yet know them. Cold latitudes are not propitious to the growth of the parasitic plant we call oïdium, and accordingly we find the more northern limits of our vine zone have thus far been most free from it. The disease has more than one form, and has been often mistaken for a mere leaf-blight by those who think themselves far beyond reach of the oïdium. New vines are generally strong enough to fling off all ailments which beset them for the first few years after they come into bearing. During those years they will commonly thrive and produce well, so that results obtained from such often lead us into error as to the value of soils, varieties, and modes of cultivation, as Mr. Sanders very well remarks in one of his late publications.

I have three vineyards of Catawbas which came into bearing in 1860, and continued to do well and showed no sign of disease until 1864, when the disease destroyed about one tenth of their fruit. The next year there was a clean sweep, and the next, and next.

As with new plantations, so it is with new varieties. The Norton's Virginia Seedling, the Concord, and the Ives Seedling are the three which have been most confidently relied on and most loudly praised for their invulnerability. My own Nortons, that had remained safe and sound since I first planted them in 1857, had the disease last year (1868), not merely in the form of "gray rot," but also in that of "*red-leaf*," the most terrible of all its manifestations. In the month of September of that same 1868, I saw, in a four-year old vineyard at M'Arthur, Vinton County, Ohio, as pretty a rotting going on as one who had foretold it would wish to see, while close beside was a vineyard of three-year old plants loaded with fruit and in perfect health. The same season, the Concords of one of my neighbors on the banks of the Ohio suffered badly and for the first time. Then, for the Ives, what meeting of the Cincinnati Horticultural Society or Wine-growers' Association has there been since August, 1868, when Mr. Howarth has not risen in his place to declare that it *does* rot, and offer to prove it? The Dianas, Rogers No. 15, and other Rogers plants, have also given way before the advancing pest, and the Delaware too.

The Catawba, that has been made a scapegoat and abandoned as hopelessly doomed, is, in fact, remarkably hardy in resisting the disease. This has been repeatedly noted in France, where it is grown experimentally. On the Lake islands and Lake shore in Ohio it withstood the invasion year after year, and fortunes were made from its fruit before it succumbed. If their Concords or Ives's hold out as long, I shall be surprised. In my opinion, the Catawba is better proof against the attempts of the destroyer than almost any variety we have while, of those whose hardiness so many have been willing to vouch for, the tonghest can only hope to be reserved for the honor of being *last* devoured.

It better behooves our vine-dressers to examine into the disease, learn the remedy, and prepare to apply it, than hug themselves in an illusory security, or fly in a panic from one variety to another, or from one place to another. But mildew, rot, and red-leaf —in other words, oïdium—can be cured and kept down. It has been done in Europe, where its march was far more rapid and sweeping than here, and here we can do it too. The evening after I arrived in Nice I bought the pamphlet of Mr. H. Marès, whom I have already mentioned as having received, at the distribution of prizes at the Paris Exhibition, a medal and a smile. I took the brochure home, and did not sleep till I had read it through. Here it is, and the reader must read it too, every word of it, that he may the better understand what is to follow it.

MANUAL

FOR THE

SULPHURING OF DISEASED VINES,

AND

RESULTS.

By H. H. Marès, Montpellier.

TABLE OF CONTENTS.

	Page
Explanation of the Figures of the Plate	214
Preface to the Third Edition	217

MANUAL FOR THE SULPHURING OF DISEASED VINES, AND RESULTS.

Use of Sulphur—its Effects	220
Development of the Vine Disease	226
Characteristics of the Vine Disease	228
The *Oidium Tuckeri*	230
Different Opinions on the Vine Disease	234
Conditions to fulfill in order to combat the Disease	236
Properties of Sulphur	236
Action of Sulphur on the Oïdium	239
Sulphuring Diseased Vines	242
Mode of scattering Sulphur on the Vines—Instruments most suitable for that purpose	247
Concerning the Epoch when Sulphur should be applied	251
Sulphuring Carignans	254
" Aramons	257
" Alicantes, Aspirans, etc.	259
" Terrets	260
Rougeau of Terrets	261
Vines Sulphured the preceding Year	264
Precepts for applying Sulphur	266
Of the quantity of Sulphur necessary	270
Of the Vegetation of Sulphured Vines	271
Review of Chapters relating to the use of Sulphur and its Action on the Vines	275
Objections against using Sulphur	277

EXPLANATION OF THE FIGURES OF THE PLATE.

Fig. 1. Reproductive spore or germ of the *Oïdium Tuckeri*, largely magnified.

Fig. 2. *Oïdium Tuckeri*, largely magnified. *m, m, m.* Creeping filaments, or mycelium. *c, c.* "*Crampons*" (clamps, anchors) of the mycelium. *t, t, t.* "*Tigelles,*" or erect filaments bearing the spores placed end to end. *s, s, s.* Spores on the "*Tigelles.*"

Fig. 3. The *Oïdium Tuckeri* in full vegetation on the skin of a grape, appearing to the naked eye as merely a white efflorescence.

Fig. 4. Spore of the *Oïdium Tuckeri* beginning to germinate.

Fig. 5. Fragment of the skin attacked by *Oïdium*, on which flour of sulphur has been scattered. *f, f, f.* Grains of flour of sulphur.

Fig. 6. *m, m, m.* Fragments of *mycelium* broken and deformed. *s, s, s,* Spores shrunk and distorted. The greater part have disappeared; only a small number are seen.

Fig. 7. Flour of sulphur, magnified.

Fig. 8. Fine *sleet* of sulphur, largely magnified.

Fig. 10. Triturated sulphur reduced to an impalpable powder, magnified to the same degree as the flour of sulphur in Fig. 7. Common triturated sulphur has nearly the same forms, but the fragments are much more voluminous.

Fig. 11. The Vergnes Bellows. *t.* Nozzle by which the air enters and is blown out again, charged with the sulphur-dust. *m.* Wire gauze with large meshes to sift the sulphur. *c.* Cavity of the bellows, serving as reservoir for the sulphur. *b.* Stopper to the orifice in the upper wood, by which the sulphur is introduced.

Fig. 12. Box with a tuft.

www.ingramcontent.com/pod-product-compliance
Lightning Source LLC
Chambersburg PA
CBHW020857230426
43666CB00008B/1218